To Anne

I hope you enjoy the excitement
And surprises I wove together
for you.

Gold Rush 2000

Ed Mitchell

SIGNED BY
AuthOR

Dust Jacket design by
Lightbourne Images of Ashland Oregon

ISBN: 0-9668447-3-4
First Edition Hard Cover

California Coast Publishing
17595 Vierra Canyon Road, Suite 407
Salinas CA 93907

Visit our web site at: http://www.goldrush2000.net
Talk to the author at: ed@goldrush2000.net

•

10 9 8 7 6 5 4 3 2 1

PRINTED IN THE UNITED STATES OF AMERICA

Tell a friend!

After you read and enjoy *Gold Rush 2000,*
please encourage a friend to buy a copy, or
purchase one for them as a gift.

Easy Order from our toll free number:
1-(888) 773-9769

Or from a book store by requesting the
book by title or ISBN.
ISBN: 0-9668447-3-4

Or send your request to:
California Coast Publishing
17595 Vierra Canyon Road, Suite 407
Salinas CA 93907

Win a One Ounce Krugerrand Gold Coin!

Prize Description:

The prize is a one ounce, Gold Krugerrand coin, 916.7 fine. Approximate value $300, depending upon spot market value on day of issue to the contest winner.

•

Submittal Information:

To win this skill contest you must do the following:

S1. Provide Proof of Purchase of *Gold Rush 2000*. In lieu of providing proof of purchase, should you receive a Best Customer Courtesy Copy, then provide the Courtesy Copy number shown on the label affixed to the dust jacket.

S2. Provide the Name of the Bookstore selling the book or provide the name of other source of purchase/gift.

S3. Provide your Name, Mailing Address, Zip Code, and Phone Number or e-mail Address.

S4. Correctly answer the five contest questions in the front of *Gold Rush 2000*.

S5. Submit all information prior to 15 December 2000.

S6. Mail entry and answers to:

California Coast Publishing
17595 Vierra Canyon Rd, Suite 407
Salinas, CA 93907.

Rules:

R1: Copies of *Gold Rush 2000* will be mailed to book stores and other purchasers no earlier than 14 April 1999. Contest begins upon first book shipment.

R2. The winner will be selected as follows:

A. Only One winner will be selected from all entries received by California Coast Publishing.
Only one Krugerrand coin will be awarded to the single selected winner.

B. Only one entry per proof of purchase of *Gold Rush 2000*. Only one entry per Best Customer Courtesy Copy number.

C. Contestant's entry must be received prior to 15 December 2000. Odds of winning depend upon number of submitted entries.

D. Contestant and purchase information must be legible and complete. California Coast Publishing and the author will not be responsible for lost, late, misdirected, mutilated or postage-due entries.

E. Answers to all five questions must be correct, as judged by the author.

F. Entries will be date stamped on date of receipt by California Coast Publishing. All entries become the property of California Coast Publishing.

G. Ties for correct answers will be broken by selecting date of earliest received entry.

H. Ties for date of earliest received entry will be broken by sending 1 to 10 additional "tie breaker" questions to the tied contestants.

R3. Winner will be notified by mail, telephone or e-mail within 48 hours of close of contest. The winner's name will be provided upon written request sent to the entry address listed above.

vi

R4. Relatives of the author, employees/relatives of the publisher, book wholesalers, and book distributors are not allowed to participate in this contest.

R5. California Coast Publishing will pay postage to deliver the prize within the 50 U.S. states. Winner agrees to pay postage for prize delivery via U.S. Postal Service outside the 50 U.S. states.

R6. By entering, winner agrees to allow use of his/her name in California Coast Publishing news releases and promotional copy without additional compensation.

R7 Void where prohibited by law. All applicable laws apply. Subject to change.

Purchase your book as early as possible for a better chance to win!

•

Questions:

Provide an answer for each of the following questions:

Q1. How many towns in Georgia are identified in *Gold Rush 2000?*

Q2. What natural and/or man-made disasters occur in *Gold Rush 2000?*

Q3. How many sparks are mentioned or identified in *Gold Rush 2000?*

Q4. Who in the Gold Association double-crosses Nolen twice?

Q5. Name the characters that Nolen kisses in *Gold Rush 2000.*

Dedication

This book is dedicated to my two mothers.

My birth mother, Elizabeth Delight Mitchell, was mentally ill and institutionalized for most of my childhood. Nevertheless, she proved to be a far better mother than some women who did not suffer the limitations that she fought each day. My mom taught me to read and to love to learn, which gave me the skills to eventually leave poverty behind and become a contributor to society. My foster mother, Hilda E. Hoseney, nurtured me through high school, helping me to heal from a difficult time in my life.

I will never be able to repay these two women for all that they did for me. May their souls rest in peace.

Prologue

By 1980, the price of gold in the United States surged above $400 an ounce, triggering a modern day gold rush in California, Nevada, Utah, and Arizona. The price eventually rose above $800 an ounce. Thousands of professional and amateur prospectors scoured the hills and streams of the West for the elusive yellow mineral. When the price plummeted below $370 an ounce, the rush evaporated.

Ten years later, micron gold sites in Nevada were investigated by geologists. A speck of micron gold is smaller than the width of a human hair. The infrequent specks are encased in rock and barely visible to the human eye. Regardless, mining conglomerates raced into Nevada to gobble up the profitable micron sites that could provide a half-ounce or more of gold per ton of ore. The 1990's Nevada gold rush was fueled by huge mining concerns who could afford to spend 100 million dollars to open a micron mine and process vast tons of rock. Today, 30 micron gold mines exist in Nevada and have already produced several billions of dollars worth of gold.

And the search continues along the west coast by men and women applying their ingenuity, determination, and courage, coupled with available technology, to find the next great strike.

The better work men do is always done under stress and at great personal cost.
—William Carlos Williams

1

Pool

Adrenaline ping-ponged through Nolen Martin's stomach as the front wheels of the pickup dropped over the lip of the ridge, jolting to a stop. He studied the faint trail twisting down the steep hill toward the roar of the waterfall. "Digger, is there an easier way down?"

"This is the only trail that stays on National Forest property, and allows us to get our half-ton of gear down to the stream," the old man answered. "Too steep on the other side."

"Like this isn't?" Nolen released his harsh grip on the steering wheel, then took a drink from his canteen to wet his suddenly dry mouth. "You sure you received the owner's approval for us to work this site?"

Digger looked out the passenger window so Nolen couldn't observe his face as he answered. "She had no complaints. Why you holding up? Can't handle it?"

"Might as well try. I never did expect to see my 36th birthday..." Within seconds, the bumps in the trail started jackhammering the pickup. Stomping the brakes did not slow their decent. "Hold on. This could be rough." Nolen wrestled the twisting steering wheel to keep the four-wheel drive pickup from flipping. It started a precarious slide sideways down the steep hillside. Fighting to avoid a boulder, he sideswiped a fir tree guarding the limit of the trail. Nolen fought

1

for control, as he bounced the pickup across the rutted bottom of the ravine, and pointed the truck up the next sharp rise. He gunned the engine, spinning the knobby tires and kicking up a shower of dry grit.

Beside him, Digger's angular frame banged about, even though the old man tried to brace his body. "Your driving ain't helping my bones any. Even Pal doesn't like this bumping around." Barking at the two men, a black Labrador struggled to remain upright in the packed truck bed.

"You were the one who told me to follow this route. Actually, I think I'm doing a...Ouch!...damn good job."

The truck flew off the top of a dirt embankment, and dropped onto a patch of powdery sand, blasting a tan cloud into the air. Nolen skidded the pickup to a halt and leaned back. "Yes!" he yelled shoving his fists above his head. "Two times, I swore we were going to roll." Nolen took off his favorite camouflaged cap, and brushed wet strands of blond hair from his forehead while watching Digger rub his elbow.

"Are you hurt?"

"I'm fine. And, I told you before to quit babying me."

Nolen shrugged. "OK. Let's inspect your idea of a great gold site." He stepped out of the pickup, then flipped the seat forward to open a protective case containing his portable computer, binoculars, and several U.S. Geologic Survey maps.

After Nolen set up the computer and maps on the hood of the pickup, the two men and the dog moved to the edge of the gravel shoreline. There, they surveyed the cold, churning creek. To their right, green water tumbled fifty feet, crashing against slate-gray boulders. From the white water pool at the base of the waterfall, the stream shot through an expanding rock canal. On the far shore, the water raced along a moss-covered, granite wall. Where the wall ended, the creek deflected left and widened, allowing the current to lose two-thirds of its momentum, and swirl past a crescent-shaped sandbar.

"You were correct when we did our library research," Nolen said. "The 1849 prospectors used the ravine as a convenient place to toss debris from the upstream mines."

Digger scratched his white stubbled chin, and squinted at the man-made dam. "Perfect! This here creek's been washing through that ore from those old claims. Even if we don't find a hard rock vein, there's bound to be gold on the bottom."

"It's perfect all right—perfectly dangerous. You want me to dive to the bottom, fight the surge from the waterfall, and poke around in the dark for gold. How long did it take you to dream up this idea?"

"Didn't take no time a'tall," the old man replied with a grin. "Just thought about how you crave adventure. Like when you were in the Army, jumping from airplanes with explosives strapped to your back, becoming a war hero by saving women and children. I knew you wouldn't back away from this site."

Nolen lifted his binoculars to inspect the rock formations near the pool, looking for fingers of white quartz in the gray stone. "Why is this spot so much better than the places we worked downstream yesterday, and the day before that?"

"Current. Here, it's fast and it's tricky. Sure to scare off most prospectors. Working other locations would be much safer."

Nolen returned to the pickup and entered site details into the computer program. He analyzed the updated map on the computer screen, with it's red speckles of clustered numbers, highlighting the ten-square-mile box they had been searching during the last week. Through some of the clusters, a black vector line sliced, predicting the presence of a gold vein. "Well, the probability program agrees with you. There's a ten percent better chance here than downstream."

"Yup, this could be the place where we find the Eastern vein of the California mother lode." Smiling and flapping his elbows the old man danced a little jig in the sand.

"Digger, I've listened to your big dream for years. All you need now is a nice little strike. One with enough gold to pay off Hilda's many medical bills, and allow you to take her to a good cancer specialist."

"Don't you forget—it's your foster mother's dream too. That's why she's stood by me for the last twenty-six years, while I've prospected both the east and west sides of the Sierra Nevada." As he spoke, Digger traced a line across the unfolded survey map with his shaky hand. "I've hunted gold from Mariposa in the south, to El Dorado in the north, one hundred twenty miles of the greatest gold country in the USA."

"Long on hope, but short on results," Nolen quipped.

"That's why over the last eight years, I've been working more to the east, in the Reno-Susanville area. It's here someplace."

"Not many promising surface signs," Nolen replied. "The original miners may have found all the lode gold in this area."

"It's been over a century and a half since then. Lots of weather and earthquakes have exposed rock that hasn't been worked yet. Remember what I told you when we studied the mining records and old maps? Any unfound veins are goin' to be in overlooked or hard to reach spots."

"Like the bottom of this pool?" Nolen placed his hand on Diggers arm to quiet him. "Hear that grinding noise? That's a boulder in the bottom of the pool shifting after the sand has been washed away from its bottom. The recent rains have got this stream raging." A shiver of apprehension shook Nolen.

"Don't you worry about the pool. After we find that branch of the mother lode, there's goin' be another gold rush in California. It'll be bigger than in 1849 when thousands of people from the east coast suffered months of hardship to come here. Just imagine 100,000 people, in their cars and trucks, flooding into this area overnight."

Nolen closed the computer and returned it to the cab of the truck. "Digger, don't forget all the problems that erupt after finding a huge strike. Remember you taught me that more people steal gold from prospectors than find it on their own."

"Stop being a pessimist, boy. That high-tech program you wrote is goin' to help find my vein. I've finally got a technology as—good if not better—than the big exploration companies! Those crooks can all kiss my ass now."

Nolen surveyed the surrounding trees for small yellow markers, used to define the boundaries of a filed claim. "Are you sure no one has staked a claim to this area? Since the price of gold rose above $500 an ounce, it's again profitable to reprocess the tailings near these old mining sites."

"Son, I checked the county records, and no one has placed a claim on this here part of the creek. So let's quit jawin', and get this truck turned around and the pumps runnin'. We've got plenty of daylight left. Move your short body."

"I've told you a hundred times, 5'8" is the optimum size for an American male," Nolen joked. "I'm not a height in-flicted scarecrow like you."

Nolen backed the truck to the edge of the embankment, where the men slid the new Keene Engineering dredging equipment and air system onto the tailgate. While Digger started the small engines, Nolen pulled on his wet suit and prepared his diving gear. He inspected every inch of the air hose that would supply him life-sustaining oxygen and its inner tube flotation support. Then, he laid out the dredging tube, underwater lamp, and safety rope, with attached com-munication cable. Nolen handed Digger a headphone, and slipped on his diving helmet to verify that the radio operated properly.

"I'll use the trailer hitch," Digger rasped into his headphone, "as the near-shore anchor point. With you hooked to the rope, I can let the current push you through the pool."

"Piece of cake," Nolen answered, hoping to hide his nervousness. "Make sure the air hose doesn't snag. I'm not so good at holding my breath." Nolen waded into the cold creek just below the waterfall, awkwardly flopping his swimming flippers, while balancing on the green moss-covered rocks. "Oh crap," Nolen swore when his foot slipped, and the current snatched him into the deep water with his safety rope and air hose snaking behind him. The stream's force rushed him toward a rock outcropping, jutting from the far granite wall.

"You're not looking real talented," Digger radioed, "floating along on your back like a flopping clown."

Nolen tried to twist his body around, to view the approaching wall, in an attempt to cushion his imminent impact. "Damn, this is going to hurt." He pancaked against the wall, slamming his left knee. Scraping one hand, he grabbed onto the jagged outcropping. He looped the safety rope around the rock for a hold against the slimy wall. It took three attempts to get braced, before he could hammer in a piton, and safety clip his rope to it.

Secured, he looked toward Digger. "This first dive, I'll try to stay about ten feet below the surface. Give me plenty of slack!"

Nolen dove, and the chilling liquid closed around him, strangling the sunlight. The torrent, ricocheting off the wall, buffeted his body, spinning and tossing him. Vertigo stalked his senses. He fought to keep his diving light aimed downstream, using its feeble shaft to warn him of any approaching danger.

Digger's distant voice echoed in Nolen's helmet. "Be careful, son. Don't get tangled in a sunken tree."

The granite wall flashed by, mere inches from Nolen's fragile mask. He tried to swim away. But, his paddling had no effect. "I'm out of control, Digger." All his muscles tightened, and he closed his eyes, waiting to crash into the rock. Moments later, the roller-coaster waves spit Nolen into the slow shallows near the sandbar, allowing him to crawl back onto land.

After lashing on extra diving weights and instructing Digger to control the drift speed by slowly letting out the rope, Nolen repeated the procedure two more times. Finally, he limped back upstream.

"Does it look good, or are we wastin' our time?" Digger pressured.

"It's like a huge vertical sieve. Some quartz marbling and many cracks in the far wall. Plenty of heavy black sand packed into the cracks. It's worth a try. However, we're not going to expose any treasure without a struggle."

"Nothing's ever worn down my will to keep trying," Digger replied.

"Your willpower, but my body's paying the price. Even though I'm in the best shape of my life, this stream battered the heck out of me. Almost broke my knee on one hit. An hour or two down there, and I'll be a mass of bruises."

"You're young; you'll recover quickly. Now, how deep is it?"

"Near the waterfall, it's about forty feet. From there, it slopes upward to six feet by the sandbar. It will be easier to handle the dredge hose, if I strap the lamp to my chest and walk upstream into the deeper water, by moving along the base of the wall. That way, you can support me by pulling the safety rope tight."

"Any obstructions?"

"Several large boulders. My worry is one might roll against my leg and pin me to the bottom."

"Son, you don't have to do this if you think it's too dangerous. I'll understand. Sending us money, and then givin' up your commission in the Army is too much already. I know it caused trouble between you and your wife."

"Our separations were not caused by paying Hilda's doctor bills. Anyway, I can't change that situation right now. What I can do is help you and Hilda."

Digger placed his gnarled hand on Nolen's shoulder. "Thanks, son..."

"It's OK," Nolen interrupted. "Let's set up the sluice box and get started."

The two men pulled a one-foot wide, six-foot long, plastic sluice box off the top of the truck and laid it parallel to the stream. Digger shoved several flat rocks under its upstream edge. Sand and water, sucked from the stream by the dredge, would wash through the sluice. Any heavy gold particles carried in the slurry, would settle behind the molded cross-slats spaced along the bottom of the box.

Nolen returned to the shallows and began edging toward the far shore. After sinking to the bottom of the stream, he studied the base of each large boulder and the cuts in the bedrock, memorizing promising places to dredge. Holding himself on the streambed became more difficult as the creek narrowed. Three-quarters of the way to the waterfall, the current's blast prevented him from taking a step without being swept backward. The current pounded him with body blows. He stopped to sink another piton into the rock wall, and hooked his rope to it for additional support.

Nolen crouched and shoved the two-inch-wide dredging tube into the sand and gravel, letting it gulp the material like a vacuum cleaner. A wire mesh on the mouth of the hose kept large rocks from being sucked up and choking the water flow. With slow motion strokes, Nolen cleaned the cracks near the base of the wall. Then, using a metal rod, he loosened sand packed within the holes in the rock slab. Routinely, he straightened his bent spine, trying to escape from the relentless pressure of the current.

Nolen began his third sweep, braced against a chair-sized boulder. "Digger, I'm looking at a triangular crack in the wall. Appears eight-feet-long. It's fist-wide at the top, then gradually spreads to about twenty-four inches across at the bottom."

"Well, quit yapping and get to work. I ain't seen any gold flakes in the sluice box yet."

"Weren't you the one who taught me to be patient when hunting gold?"

"Tossing handfuls of worthless pebbles back into the stream isn't helping my Hilda any," Digger grumbled.

After clearing the narrow upper portions of the crack, Nolen squirmed onto his stomach. He swung the hose back and forth along the base of the wall, excavating a shoulder-wide cavity. He aimed his diving light into the opening. Thousands of brown and tan specs swirled in the lamp light.

"I think there's some glitter in the back of this hole I'm working. I can't tell for sure."

"Maybe Mother Nature is ready to give us something of value," Digger answered rubbing his hands together.

"Or, just luring me into a trap. Only way to find out is to squeeze farther into this damn crevice." Nolen crawled forward and shoved his arm and the hose deeper inside. Sand slithered into the tube, exposing the top of a nugget. He stretched his arm, but his fingers wiggled long inches from the gold.

Nolen backed out and wedged the dredge hose between the wall and a nearby boulder. "I couldn't reach it, Digger. I'll have to go in farther."

"Will your claustrophobia let you enter the hole?"

"Ever since I was a kid I've told you I'm not claustropho-
bic." Nolen tried ignoring the sudden clammy sweat inside
his suit as he unhooked his diving lamp from his chest. With
the lamp and hose in one hand, he re-entered the narrow
crack. Using his elbows and toes, he inched forward. When
the hard rock bit into his sides, a shiver skidded up his spine.
Too damn tight in here, his mind shouted! Compulsively he
scurried out, his breathing rate tripled.

"How are you doing, boy?"

"Leave me alone, damn it. I'm busy." Attempting to regain self-control, he closed his eyes, then counted to twenty
as he slowly re-entered the hole. To prevent his mind from
focusing only on the rock's firm body grasp, he concentrated
on the yellow sparkle, deep within the crevice. He scolded
himself for panicking. Quit being a wimp, he thought. You
can't give up on your promise to help Hilda, just because it's
getting tough.

"Yahoo!" Digger cheered. "Eight dime-sized nuggets just
settled behind one baffle. You're into a good pocket."

Encouraged, Nolen squeezed farther into the hole.

"You don't need to point the gun at me," Digger rasped.
"I won't do nothin' foolish."

Nolen stopped dredging. "Digger, who the hell are you
talking to? Why is Pal barking?"

"Son, there's some young woman up here, pointing a
shotgun at me. Don't shoot lady! I've got to get my boy out of
the pool!"

Through the earphones, Nolen heard two muffled explosions. Then, he felt the sudden drop in air pressure inside his
face mask. Frightened, he gulped air before the ice cold water shocked his face. He drove his hands into the sand,
scrambling to push out of the hole. But his weight belt
jammed against the rock. The need for oxygen began claw-
ing at his chest and throat. Nolen twisted and squirmed
backward with all his might, ignoring the pain as the rock

and belt ripped his skin. He popped out of the hole. Tearing off his weight belt, he struggled to keep his mouth shut to avoid sucking in suffocating water.

After what seemed an eternity, Nolen surfaced, coughing and spitting. He fell to his hands and knees, on the sandbar gulping air. Looking up, he spotted the ugly black eye of a pump shotgun being held by the woman. She stood, partially hidden, behind the trunk of a nearby fallen redwood. The tree and a bulky coat hid her features, except the mop of dark curls surrounding her face and the determination flashing from her eyes.

"What's...what's going on," Nolen coughed.

"Get off my land!" she shouted, then aimed slightly behind him.

The roar of the shotgun told Nolen all he needed to know. He bolted for the pickup with his flippered feet scattering sand and water. Jumping onto the running board, he lunged for the side mirror as Digger stomped the gas pedal, fishtailing the pickup.

Through the sudden dust cloud, the woman fired again at Nolen.

2

Terminations

T.J. Tower stood gazing through the window of the executive suite on the twenty-first floor of the Continental Mining and Refining skyscraper. He paid little attention to the impressive morning view of San Francisco, as he listened to the whispering mood of the people entering the adjacent conference room. He knew the men and women settling around the long table were observing him through the open double doors, trying to gauge the stranger who arrived after the hostile takeover by the Japanese commodity company.

Mentally, they compared the fifty-three-year-old outsider to their trusted boss seated beside Tower's desk. They studied the newcomer's tailored suit, his aloof stance, and his tanned features. Each executive watched for any signal which might reveal what impact this man would have on their future.

"Earl, close the doors," Tower commanded.

Tower's nephew, who had accompanied him from Japan, jumped to the task.

Tower faced the Vice-President for CMR's Pacific Division. "You're fired. You have one hour to clear out your desk."

The Vice-President dropped his cup of coffee. "You son-of-a-bitch, you don't have to let me go. You could shift me to a special assistant position, so I can retire with some dignity."

"Security is already at your office. We've confiscated your computer and disks. Your private files will be turned over to you in a week," Tower barked.

The stunned Vice-President slowly shook his head. "What we learned about you during the takeover battle was right. Now, I know why they call you Mr. Garbage."

Tower clenched his teeth. "You're like all the miners I've ever met, from Syria to Australia to the U.S.A. You make horrendous mistakes, then hate me when I'm sent in to clean up your crap."

"Don't try to act righteous," the V.P. answered. "You enjoy being Tokyo's hatchet man. And no matter how much you try, you'll never be accepted by anyone with any decency."

"Earl, have Security step inside and escort the ex-Vice-President to his office, now!"

The V.P. shoved his chair away from the table and stood. "Even though you graduated from college, changed your name, and lost your drawl, you will always be white trash from Louisiana. Your veneer doesn't mask that fact."

Tower's ego bristled. "Despite what you think, only a few of your peers will be retained. Those few will be allowed to stay because they possess some talent and intelligence."

"So you can use them for your own benefit, I suppose!"

"Of course, with most of you out of the way, I have a clear path to become President of CMR. You'll see me gain the influence, respect, and prestige I deserve, as I move ahead of you and those other hypocrites."

"It's not as easy as you think, Mr. Garbage. Now, you have to deliver. And your past acquisitions invariably have had a stink surrounding them. That type of behavior will hurt you here."

"Why should I worry? This isn't the first time I've dealt with mineral properties. I know more about mining law than ninety percent of the lawyers I battle. That's why I've made a profit on every property or commodity I've handled for Hokido International."

"Like the weapons you sold to the Mid-East terrorists?"

"There's no proof I ever dealt with terrorists." Tower fingered a two-ounce gold nugget on his vest chain. "I can't stand you sanctimonious miners. Even so, every miner I've ever dealt with has walked away with more than he deserved."

"You have the ethics of a rat. You think screwing trusting people is what they 'deserve'."

Tower gave the executive a smug smile when the large security guards entered the office. "Your escorts are here. Have a nice day."

After the angry manager was led away, Tower entered the conference room and strode to the head of the mahogany table. His stone-hard glare shifted from face to face, boring in, demanding silence. "During the last few days, I've evaluated this division's five-year exploration performance. I've read your reports—several with faulty conclusions. I've listened to your briefings, studied your capabilities, and uncovered several weaknesses. Bottom line—most of you need the obvious explained. For example, why the previous Division Vice-President was fired."

The senior executives seated around the long conference table shifted nervously, struggling to control their concerns. Startled glances volleyed between the less experienced mid-level managers seated along the side walls and at the rear of the room.

"I sacked him because this eight-state division is not producing—which means you are not producing! I'm here to correct these problems, and return CMR to a reckoning force in the exploration market."

A chubby man at the large table spoke. "T.J., that's a little harsh. Many of us have been here for years, and done quite well."

"Excuse me," Tower interrupted. "I'll let you know when to refer to me as T.J. Until then, Mr. Harrison, address me as 'Mr. Tower'."

Crimson crept up the fat man's neck. "Ah, yes," he replied.

"Mr. Tower," a female department head interjected, "most of our profit reduction is due to the drastic drop in the output of our oil reserves."

"Wrong! Furthermore, your feeble excuse simply reflects a major problem within this division. Our profit loss was the direct result of how this division has been allowed to do business. You encouraged corporate management to depend upon CMR's 1990 through 1995 Alaskan oil discoveries, and the subsequent high oil profits, to carry this company. Meanwhile, there was no sustained hard metal development. That strategy made your jobs easy. But now we are reaping the benefit of that poor strategy, and the resultant slashed cash flow when those oil pools turned out much smaller than predicted. Our common stock value has eroded. It is down thirty-two percent.

"Simultaneously, this division reduced its research budget by ten percent a year, while our competition is using new technology to make a fortune. Isn't that correct, Mr. Pitt?"

Surprised, a balding engineer, seated against the wall, jerked when his name was called. "Ah, pardon me?"

"Mr. Pitt, don't you supervise CMR's research and development efforts?"

"I did. I mean, I do. Yes, sir."

"Please enlighten everyone as to what our competition, Homestead Mining, has accomplished during the last two years."

"I suppose, uh, sir, you're referring to the extraction process which they just implemented at Indian Flats in Nevada. Homestead's reclamation technique is reputedly 300 percent more efficient than any other American smelter. So, they are now able to wash gold from ore, previously discarded as unprofitable. But sir, I must say, during the last four years, they committed over 100 million dollars a year toward developing their technique. It's not a risk-free endeavor."

"However," Tower responded, "now that the price of gold has risen, Homestead will pay off all their sunk costs this year. Many market analysts predict Homestead will also rake in an additional fifty million dollar profit. Their long-range strategy, which included a willingness to accept risk, is putting them far ahead of CMR. So, Mr. Pitt, just how long would it take us to develop a similar technique?"

"I would say probably two, maybe three years, Mr. Tower."

"Have you begun working on a such a process?"

"Ah, not extensively, sir. No major upgrades have been instituted in our smelters since 1996."

"Have you considered hiring a Homestead employee, one who has been working on their technology?"

"Of course not! We've never resorted to..."

"Well, it's time we think about it, Mr. Pitt. Your deadline for a plan is tomorrow. After I hear your report, I'll decide whether or not you still have your job. And since you mentioned smelters, that brings me back to Mr. Harrison. What about the one you oversee in Newberg, Nevada?"

The jowled executive apprehensively fiddled with a pencil while replying. "I'm sure you're aware, it's the oldest facility in CMR. We've cut the work force to a bare minimum. We plan to renovate it when CMR has available cash. Or, when we discover another major mineral strike."

"What you failed to tell us, Mr. Harrison, is that you are operating the most costly smelter in the West. You did not elaborate why you cannot rapidly turn it into a positive profit center. Do you think I'm unable to recognize such conditions? When do you plan to close the smelter?"

The executive winced. "Mr. Tower, I can't believe you would consider that action. Our facility is the major employer for Newberg. Those eighty dedicated families have supported CMR for six decades."

"The formula is simple. We protect the remaining 833 people in this division, by sacrificing the Newberg group."

"But the company is not losing money. Our profit margin is merely much tighter. Closing the smelter would be a disaster to CMR families in that town. Is that really what you want?" Mr. Harrison challenged.

"I want whatever it takes to end stagnation in this division. If that means closing the smelter, I'll do it. Can someone explain why CMR isn't using satellite remote sensing to locate favorable exploration sites with major fracture systems or even reprocessing tailings like other mining conglomerates?"

"We rejected using satellites," Mr. Pitt answered, "because the marginal return appeared too low to justify the cost."

"Oh, bullshit, Pitt! I need you people to realize this division is performing just like the old American auto and steel industries. You are fat, comfortable, and being left behind by hungrier, more efficient competitors. Well, I can assure you that I'm here to end any complacent attitudes, and I intend to turn this division around—by whatever means it takes."

"How do you propose to do that?" the female department head questioned.

Tower snapped his fingers. "Earl." Tower's administrative assistant began handing each person a three-page pamphlet.

"First, the old ways of doing business are over—finished—done. Forget the massive binder of policy letters that have bogged down this division in bureaucracy. In your hands, you now hold the new division policy. It is short and simple. Results count. Achieve them any legal way you can. And those who can't produce, won't be here much longer."

For emphasis Tower made eye contact, one at a time, with each of the senior executives surrounding the table. "Secondly, state supervisors will leave their plush offices at least one week a month. Each of you will travel to the field and talk with all your exploration teams. I expect this head-quarters to recapture the lust to discover something other than paper reports."

A junior executive spoke from the back of the room. "We could provide a cash incentive to wildcatters and independent prospectors."

All heads in the room turned to stare at the unexpected speaker. "Well," he continued, "hundreds of new prospec-tors are out there, since the price of gold jumped. Why not tap that reservoir of enthusiasm?"

"We've tried grubstaking before," Harrison declared. "It did not work."

Tower pointed to the young manager. "You—see me in my office, later. And Harrison, he said 'incentive', not 'grubstake'. Furthermore, this young man is the first within this division to offer a positive suggestion, instead of telling me why we can't do things differently. His plan just might entice some ambitious men to join with us to produce re-sults. Which, I might add, is certainly what this division needs."

"You mean men and women," the female executive added.

Tower grunted. "Excuse me, I forgot I'm in California. You are exactly right. I do not care about gender, just about results. Prove to me you can find gold, silver, gas, oil, or mo-lybdenum, and you get promoted. Did I express myself better?"

"Fine," the woman replied.

"All right," Tower continued, "you have four days until the next conference. When we meet again, each state supervisor will brief me on ways to slash costs and increase profits. I want a list of feasible alternatives. Be ruthless. Now, get going."

As the men and women began filing from the conference room, Tower signaled his assistant to follow him into the executive office. "Are the dinner arrangements completed?" Tower asked.

"Yes, Uncle. Earlier today, I spoke with the three Vice-Presidents in locations away from their offices. None of our competitors will suspect that you are attempting to hire their best talent from them."

"What about the hostesses?"

"Everything is settled. I paid cash for the most sophisticated prostitutes this town can offer. These ladies will make your guests believe their own wit and charm enticed them into bed. Even if the executives do recognize them as working girls, the women are beautiful enough to cause the men to overlook any moral concerns."

"Did you select one for me? I don't want to scare off these managers."

"Yes. And I told your wife you would be working late tonight, planning the fund raiser for San Francisco's homeless. By the way, the Mayor called to thank you for offering to host the charity event."

"Good excuse. And the video camera systems?"

"They are in place and operational. You will have plenty of film to encourage any hesitant VP to join your staff, before firing any of the present CMR managers. Which is wise, since it appears you have already identified who you plan to terminate."

"Yes, I set up several of the fat slobs during this meeting. The last action they will handle for CMR is telling me what we should do to be more efficient. I'll claim they were not imaginative enough, then fire them. Not only will I get rid of the weak personnel, while cutting my overhead costs, but I'll also show those remaining that I do mean business."

"Another fine start, Uncle."

At his desk, Tower shook his head, concerned, as he studied a map detailing the three dozen producing mines established in the last eight years in Nevada and California. For the tenth time, he noted that CMR's competition had outmaneuvered them in Nevada's gold fields during the mid-1990s, when CMR was too slow to acquire a significant, profitable gold mine. Now the competition was flush with money. And in every herd there is a weakling which is eventually cut out and killed. CMR was the weak one now.

"Don't be fooled, Earl. Kicking ass is easy. But locating valuable minerals is tough. And the Hokido Board of Directors gave me only one year to turn this division around; or else, next year you will see me being canned."

Was the fired exec right, Tower thought to himself? Maybe the Jap Board of Directors hate my white guts like the people in this company now do? Are they just using this assignment to justify firing me because they know it is near-impossible to turn the situation around in a year's time? No down-side for the Directors. If I turn it around they win. If I can't, they fire me. Screw the bastards! I can always help the competition absorb CMR. There's always options.

3

Maida

5 January

Nolen jerked his feet through the truck's window and fought to right himself, as his head repeatedly smacked the floor of the bouncing cab while Pal kept barking from the truck bed.

"You all right, boy?" Digger shouted.

"All right? I damn near drowned in that hole I crawled inside. A crack so tight I couldn't expand my chest to even inhale. Suddenly, my mouth is full of water. My back is torn open, grappling to escape. Then, when I reach the surface, Annie Oakley tries to blow a hole in my ass because of you. Hell yes, I'm just dandy!"

The old miner downshifted to keep the truck roaring up the incline. "Well, I didn't mean for things to turn out this way."

"You didn't get the property owner's permission—did you!"

"Not exactly," Digger whispered, ashamed.

"You've never done anything this stupid before. Why now?"

"Because I need money, right away. Hilda must return to the hospital—treatment for more tumors. Besides, I did mail an offer to the owner of record, probably that girl. I just didn't wait for an answer. I was hoping to avoid all the hagglin' that slows start-up operations. I didn't think she'd be up here. The survey map doesn't show any buildings near the falls."

"Dumb ass decision!" Nolen growled. "We're lucky that gal didn't kill us."

"I'm...I'm sorry. You know I would have shared any findings with the landowner."

Nolen peeled two egg-sized nuggets from a pocket in his wet suit. "We darn will share! Because this gold doesn't belong to us."

"Well, son, I told you it was a perfect spot."

Afternoon sunlight filtered through the pine and fir trees, as Nolen limped along the muddy road leading toward the lake. He held high a long stick, dangling a white T-shirt. As he stepped from the tree line, he spotted a small cabin, probably built after the U.S. Geologic map had been published ten years earlier. The building sat, mere yards from the National Forest boundary, near the edge of the lake, facing downstream toward the top of the waterfall. A compact car was parked next to the front steps.

The door slammed open. Threatening with her shotgun, the dark-haired woman stepped onto the porch. "I warned you once. I'm not letting you ruin my property. I'm calling the police!"

"Wait! I brought what I extracted from the bottom of the pool. There's about two pounds of gold here. It's worth $12,000 or more." Nolen held up the large, shiny nuggets.

Her uneasy expression warned Nolen that she did not believe him.

"I'm sorry about what happened," Nolen continued. "I thought we had approval to work the pool."

"I mailed a reply last Wednesday, emphatically stating no," she declared.

"We were already in the field by then. When I learned the truth, a few minutes ago, I knew we must return your gold.

I'll leave the nuggets here. All we want to do is collect the gear we left at the creek. We'll drop off any gold currently in the sluice when we leave."

"Why didn't you run off with the two nuggets? I doubt your equipment is worth one-third of what you discovered."

"Look, we're not dishonest. My foster-father just made a bad decision. His wife's sick, and he needs money for her care. I'm just attempting to help him correct his mistake. OK?"

"Put the damn gold on the porch, buster," she ordered. "Then move away."

Nolen studied the woman's sinewy, athletic body under her leotard as he walked forward. Darn ironic, he thought. The first pretty woman I notice in months has to be pointing a shotgun at me. He set the nuggets on the edge of the deck and stepped back.

"Last week," she said, "the local sheriff told me a man was found murdered on his property a few miles north of here. So, I'm going to stay cautious until my two brothers return. They are due back any moment."

"I take it we can retrieve our mining gear?"

The woman picked up one of the nuggets while trying to keep her eyes on Nolen. "Heavy," she exclaimed, surprised.

"Never held raw gold before, have you, lady?" Nolen noticed her repeatedly drifting a look at the remarkable nugget in her hand.

"Tell me," she inquired, "is there more gold in the pool?"

"When you find large nuggets, usually there's flake gold. Sometimes, quite a bit."

As he started to leave, the woman called out. "I have an offer that might benefit us both."

Nolen turned and waited.

"I'll hire you to dredge the pool for fifteen percent of what we find. Your dad will earn some gold to take care of his wife. And my brothers and I won't have to extract the gold ourselves. Deal?"

"It's dangerous at the bottom of that pool," Nolen replied. "I should get at least one-third for risking my life."

"That's ridiculous! It's my...ah...our gold."

"Well, only one vehicle, presumably yours, made those tire tracks through the mud along the cabin driveway. And there's no sign of any mining equipment. I'd say you don't have any brothers, nor any clue how to mine gold."

"You're not as slow as you look," she snapped. "But I'm not forced to use your services to retrieve the gold. I can easily locate someone else knowledgeable about dredging. The sheriff said recently many prospectors have arrived in the county."

"You're right, as long as you can finish soon. Because today, I disturbed the gravel and sand packed into the rock wall. The gold is now exposed to the current and will quickly begin to wash away. After you show someone else the dredge site, I doubt they'll do the work for less than I'm asking. At least, not anyone who won't cheat you. Lady, I've already proved that I'm honest."

"Somewhat honest," she conceded. "I'll give you twenty percent."

"Make it twenty-five, and I'll explain how we do the work, allowing you to watch us, to ensure we don't take more than our fair share."

"I'll only agree to twenty-two percent. And I still don't trust you. Put your driver's license and a credit card on the porch, then step back again. I'm going to call the sheriff and give him your description. He can verify whether or not you have a criminal record, before we start dredging my pool."

Nolen smiled, appreciating her business savvy. "It's a deal. My name's Nolen Martin. What's yours?"

"Maida Collins."

"Miss Collins, be careful what you say to the sheriff. We don't want any rumors started that large nuggets have been found. Otherwise, you'll have people crawling all over your property, bringing with them more trouble than you ever dreamed possible."

She hesitated, unsure after Nolen's caution. Her eyes flirted back and forth between Nolen and the nugget. Finally she said, "Move back another thirty feet while I use my cell phone."

Later, after Digger had mumbled his embarrassed apology to Maida, Nolen repaired the air hose, then dove into the pool. He worked underwater, fighting the swirling current, until his legs began trembling with fatigue, forcing him to surface for short rests. He tried staying under for half an hour at a time. But, as his strength faded, he waded ashore more frequently.

Meanwhile, Digger transferred the black sand, accumulating behind each baffle of the sluice box, into a five-gallon plastic bucket. When the suction tube coughed short flashing spurts of nuggets, Digger plucked them from the sluice and dropped them into a coffee can placed next to Maida.

She watched, occasionally laughing, fascinated by the unexpected wealth spitting from the mouth of the dredge. She selected a pea sized nugget from the sluice and held it over her right hand. "Wouldn't the girls at work die of envy if I walked in with this on my finger surrounded by some blue-white diamonds? It's amazing that it's free and coming out of my pool!"

Nolen flopped onto the sand next to her. "Free isn't quite the right word," he panted.

Maida turned to Nolen, as he lay on the bank massaging his cramping thighs and calves. "What do we do with the black sand that Digger is collecting?"

"The sluice box is only the crudest separation step. It washes away the lightweight brown dirt and sticks from the heavier black sand and gold, if there is any. Next, we use plastic pans to carefully hand wash the black stuff into a metal tub, leaving visible flake and nugget gold in the pan. Back home, we process the black sand a third time, using a smelting technique where mercury captures the very smallest specks of gold."

"I own twelve acres along the lake and stream," she exclaimed. "I could own tons of gold. Quick, teach me how to do the panning."

Nolen sat Maida next to the stream with a miner's pan in her hands, and instructed her to mimic his actions. She scooped a handful of ore-rich sand from the bucket into her plastic pan. Then she began sloshing the mixture back and forth in the stream. A brown stain floated into the water. It exasperated Maida, that her panning was clumsy and took far longer than Nolen's expert movements. Finally, she reduced the slurry to a three-inch, black triangle, piled halfway up the wall of her green pan. Following his instructions, she slowly backwashed the material. "Look, look, look!" she shouted, tugging excitedly on his sleeve. "It's so beautiful. At first there was only black sand. I thought there wouldn't be any gold in my pan. Then it popped out—just like that! Now, what do I do?"

Nolen handed her an eye dropper. "Push the air out of the eye dropper, before you put it into the water in the pan, or you'll blow the gold flakes away from you. Then suck up the gold without getting a bunch of sand with it."

Maida repeatedly squirted the metal chips that she captured into a clear plastic jug. The small bits began powdering the bottom of the container, forming a yellow pyramid, rising toward the top. She grinned at Nolen. "This is fun! Now, you've had enough rest. Digger and I will keep panning while you dredge some more gold."

"Great. Now I'm working for two slave drivers."

Long shadows had stretched across the creekbed by the time Nolen wearily crawled onto shore, to disconnect his safety line. Digger shut off the dredge and air pumps, then handed the coffee can, containing small nuggets, to Maida.

"Hurry—get dressed," she encouraged Nolen. "So we can go to my cabin and weigh the gold, as you promised."

Nolen looked at her, then to his left and right. "You want me to take off my wet suit in front of you, or do I get some privacy?"

She spun around, and strode toward the campfire Digger was igniting. Nolen stripped the steamy, black rubber shirt and pants from his tired body, and tossed them into the pickup. He stood nude, at the edge of the stream with his back toward her, using the cold water to wash his sweaty body.

Nolen studied her reflection, shimmering in the side window of the pickup. Those high cheekbones and expressive eyes give her a distinctive look, he thought. Good enough to make other women stare. Good enough to make me wish.

He watched her mirrored image glance over her shoulder, as he toweled himself dry. When Maida realized that Nolen saw her staring at his body, she quickly averted her eyes. But her resistance soon failed, and she resumed furtively studying him.

After Nolen pulled on his clothes, the two began hiking to her cabin. "Is mining what you do for a living?" Maida queried.

"No, Digger's the prospector. He and his wife, Hilda, took me in when I was eleven and raised me. I'm just helping until he pays Hilda's medical bills, and gets back on his feet. Actually, for the past fifteen years, I've been an anti-terrorist."

Maida laughed. "Yes, and I am the President's closest advisor."

"Don't look at me like I'm strange. Honest, I'm adept at saving world leaders and beautiful women like you."

"Exaggeration appears to be one of your character flaws," she observed. "Let's return to reality. How did you choose my pool to dredge? Do you just drive around until you find a spot you want to plunder?"

Nolen smiled. "No. It's taken us six months to prepare for this first trip. I developed a computer program to statistically predict gold vein locations. Then we spent weeks developing the input database. We combined Digger's knowledge about conditions at thousands of sites across the Sierra Divide with facts taken from historical mining records."

"I was not aware prospecting was accomplished with computers."

"Large mining and oil companies commonly use computer analysis. But individuals like Digger rarely have the money to develop the search programs, or the expertise to use them. Some remarkable searches have been accomplished with that technology."

"Give me an example."

"Well, in 1974, the CIA and the Navy used a program to help find a Russian submarine that sank in the Pacific. Our government wanted to obtain the Soviet communication code books from the sunken wreckage."

"How did they use the program?"

"The Navy knew roughly the surface location of the sub before it went down. The program allowed them to identify the most likely places to search, within a fifty-square-mile region. Their salvage ship swept magnetic meters and video cameras across those sites. After a couple of attempts, they hit pay dirt. Then they used a mechanical claw to retrieve the portions of the sub that they wanted."

"How does the computer search technique help find gold?"

"Just like the Navy, first we identify the best places to look, then sample the site. When we spot indications of gold, we use special equipment to extract the ore."

"On your own, you developed a search program, similar to that of the Navy? You must be a computer whiz."

"Not really. With the Freedom of Information Act, I was able to uncover a copy of the program. I even located and spoke with a professor from the team that developed the original code implementing the mathematics. Since I didn't have to start from scratch, it only took a few months to modify and debug the code."

"So, the gold we found proves your program works."

"It's too early to tell. Our objective is to find hard rock veins, the source of loose gold found at the bottom of streams and pools. This is our first field application, and we haven't found a vein yet. Maybe we were just lucky today."

They entered the cabin to begin weighing the gold on the small scale Nolen had brought. The day's dredging produced fifty-one ounces of loose gold and the two egg-sized nuggets.

Maida placed a steaming cup of coffee in front of Nolen, then sat next to him at the oak-plank kitchen table. "Do you expect we'll do amazingly well again tomorrow?"

Nolen sagged in his chair, attempting to stay awake and answer her question. "Based on all that Digger's taught me, and my study of geology, I doubt it. Today, I vacuumed most of the cracks in the rock wall. Tomorrow, we might find some glitter in the downstream sandbar. However, little may be buried there, because it gets washed away and rebuilt several times each year."

"What about locating a vein?"

"When we quit finding loose gold, we'll start the tough, slow job of digging out some of the quartz. Then we'll test it for mineral content. However, I don't think there is a vein here."

"After finding so much gold today, why are you so pessimistic?"

"Never expect too much when you're hunting gold. It will seduce you into wasting all your time, energy, and money, chasing a false fantasy. But enjoy your luck. Days like this are truly rare."

"And wonderful!" she exclaimed. "For seven years, I have been selling stocks and bonds to people in San Francisco, scrambling to earn small commissions. But during one short day in my backyard, we find nearly $42,000 worth of gold. That's nearly as much as I make in a year. Now, I can send a check home to help pay the farm mortgage that my brothers have been struggling with."

Nolen rubbed his aching knee. "The unexpected money should make them pleased with you."

As was her habit, Maida chatted while curling a lock of her hair around her finger. "My parents are proud of me, but they do nag. They feel I haven't earned enough money to warrant being so far from home. But then, they can't ever say how much is enough. Constantly, they remind me that the big city is not the proper environment for a single girl, with morals and values. They want me to return to Georgia to marry a nice local boy."

"Will you?"

"I do miss my family and the farm. Often, I recall fond memories of my childhood. It's so peaceful there...so different from my present lifestyle. I'm simply not sure."

"Sounds like some of the country remains in the country girl."

Maida giggled. "You can run fast. A bit awkward, but fast."

"Most people can break world records with someone firing a shotgun at them. Why did you feel it necessary to shoot apart my air hose?"

"Sorry. I fell in love with this place when I first moved to California. Every other week, I work here on my computer, instead of in the city. My little place is ideal to exercise, read poetry, and enjoy the lake scenery. I was jogging past the top of the dam when I saw two strangers trespassing on my property. I felt violated. I thought you had deliberately ignored my letter—making me so angry. After retrieving my shotgun from the cabin, I decided shooting first would scare you enough to make you leave."

Nolen nodded. "Your strategy worked."

Maida rested her hand on Nolen's forearm. "I want to tell you, I'm grateful you returned, and also for your determination to overcome the current in the pool. Brawn, brains, honesty, and the ability to find gold. A remarkable combination of qualities. Thanks."

Nolen enjoyed the warm touch of her fingers. She had not bathed yet and he was acutely aware of, and relished, the musk smell of her perfume mingled with her sweat from the day's work. "No need for thanks. It's your land and your gold."

"Still, you didn't have to return the two nuggets to me. Instead, you could have monitored my weekly routine, selected a time when I was gone, and simply helped yourself to all the gold in the pool."

"Not my style. I know several of Digger's prospector pals. They believe that once they get their hands on enough of the metal, all their dreams will come true, and they'll be happy forever. When that intoxication seizes you, it's difficult to do what's right. I learned long ago, wealth doesn't guarantee happiness."

"What is the key then?"

"For you, maybe living on a Georgia farm with that boy."

Maida laughed. "I'll take my chances here. And thanks anyway, for the gold and for the advice."

"Fine. Someday you can repay me. But right now I'm going back to camp for some long overdue sleep."

She hesitated, glancing at Nolen with her dark eyes. "You and Digger don't have to sleep in the cold. You're welcome to bring your sleeping bags to the cabin."

Nolen smiled. "What—and get shot when I start stumbling around trying to find the bathroom in a dark, unfamiliar cabin? No way. Anyway, I'm so tired I doubt I could make it up the hill again. So, not tonight, Miss Collins."

"Please call me Maida."

Nolen laughed as he stood up and gingerly stretched his painful muscles. "Great, I'm one of the good guys again."

4

Memories

Nolen slept soundly in his sleeping bag until the disturbing dream of his tenth birthday returned...

He had walked home late from grade school trying to delay the inevitable closeness, triggering his stepmother to curse at him throughout the afternoon. When his father traveled on business trips, being ridiculed by the woman was far better than being around her for an extra minute. And since she only yelled at him while he completed the evening chores, his plan was working. It would have been safer though, if his older brother hadn't run away a month earlier.

Knowing how unpredictable she could be, he remained extremely careful as he sat down at the dinner table. He edged his slight frame as far from her as possible. And he avoided doing anything to rile her emotions, like looking at her face or even speaking. But his hidden nervousness betrayed him when he started to raise his glass of milk. It slipped, banging back onto the table, splashing a small drop of white liquid onto the plastic tablecloth. He tried to apologize. Ignoring his pleas, she backhanded him off his chair, then sent him to bed without allowing him to finish eating.

Now, in the small, dark bedroom, his radar-sharp senses snapped alert, warning him of the approaching danger. Her slippers swished across the hallway rug, as she walked toward his bedroom. Quickly and quietly, he curled into his best protective position. He turned his shoulder blades toward the door, and slid the pillow over his head. He pulled into a tight fetal position using his legs and arms to cover his face, stomach, and groin. He clenched his hands, to ensure his fingers were not exposed, then listened to her ragged breathing, coming from the other side of his door.

What did I do, he thought? Why can't I please her?

To gain control of his shaking body, he concentrated on three phrases, repeating them like a mantra.

No one loves me.

Think and stay calm.

Fight to survive!

When he heard the doorknob turn, his buttocks tightened, and a hollow, sick feeling washed through his stomach.

"What are you doing, you little bastard?" she snarled. "Trying to irritate me again? You want to keep me awake all night, don't you? You like to squeak the bed springs, just to bother me. Well, maybe this will teach you!" She raised a cut-off broom handle, and slammed it into his blanket-covered form.

He hated to cry in her presence, because she enjoyed that. He had learned how to tolerate the pain of the beatings without showing the hurt. But after the sixth blow, he knew she was more angry than usual, and might break one of his bones. He started hollering and crying, letting her know she was hurting him. He knew well that his pleading usually helped to satisfy her desire to strike him.

"Don't yell, you turd. You can't fool me. You're just trying to make the old couple next door hear your ruckus. So they'll get the social worker to drag us into court again. You always embarrass the whole family. I knew when I married your father that you were bad. I warned him you would cause trouble. It's the devil's blood running through your veins that makes you evil! You'll either mind me, or I'll give you away some day. Do you understand me?" She clubbed him again, and again, and again.

He squirmed away, pulling the pillow along to shield his head, while attempting to guess where the next blow might strike. He tried to withstand the hot stinging attack by moving just enough to lessen the impact from each blow.

"You little coward. Jumping around won't help you!" She yanked the covers away and chopped down with the sinister stick, crashing it against his exposed elbow. Hot pain rippled through him. Numbness replaced all feeling in his arm. Tearful anger muted his fear. He must do something to resist, or she might cripple him. He threw the pillow in her face and yelled, as he lunged to stop the club.

Nolen jerked upright from his sleeping bag, a snarl slashing his face, with fists clenched so tightly his knuckles ached. The shock of the nightmare remained with him.

Looking around, he detected Digger his weathered, lanky frame stopped next to the morning campfire. The white-haired man dropped a log onto the coals, kicking a cloud of firefly sparks into the air. He spoke in his usual raspy voice. "Boy, you sure spook Pal when you jabber in your sleep and wake up hollering."

From behind the old man's thin legs the Labrador, solid black except for two white front paws, eyed the younger man.

Cold morning air washed over Nolen's sweaty chest. A shiver shook his body. It was caused more from the vivid dream, than from the winter breeze blowing across the snow-covered Sierra Nevada mountains. I'll never let that woman win, he thought. He stretched his arm toward the glowing fire, and plucked a red ember from the ashes. He squeezed the tiny coal between his thumb and forefinger, inviting the searing pain to dance through his body. Slowly, the frightening memories slipped into the dungeon recesses of his mind.

Nolen dropped the darkened ember onto the ground and studied his burnt fingers. Satisfied that he had returned to the reality of the moment, he reached for his woolen shirt and slipped it around his muscular torso. Sweat dripped from his face as he ran his uninjured hand through his blond hair, attempting to straighten the tangles.

Digger squatted, warming his gnarled hands over the growing blaze. "Its been about twenty-four years since I heard you yell in your sleep. You did it almost every night for months after coming to live with Hilda and me. One of us sat beside you each evening to show you no one would hurt you. I thought you had gotten over those fears."

"Self-control doesn't completely overcome some things. Certain memories are never forgotten, just buried." Nolen pulled several boxes from the first aid kit and began medicating his burns.

"Boy, you always come out of those dreams real angry. Reminds me of when you were in high school. Every other week, I'd have to pull you off a bigger kid."

"Sometimes, a person is forced to fight."

"Well, I'm glad I never had to handle you when you're riled. You snarl and slash like a cornered wolverine, who'd rather die than lose."

"You know I'm easy-tempered, till someone starts shoving me around."

"Yeah, easy-tempered like a smoldering volcano. Must be something pretty strong causing you to recall those bad times?"

"Probably the bruises all over my body from yesterday's dive. Maybe the recent separation from my wife. Or hearing night noises, similar to the sounds that scared me when I was a kid. Like your stumbling around during my sleep."

"Been worrying again, if your past is messing up how you treat people? By now, you should know you're normal. You didn't turn into a delinquent. You were commissioned, and then decorated as a hero for saving mothers and children, during the war with Iraq. Even got a degree in computer science. What's it going to take for you to let go of your doubts?"

"Might be nice to have one relationship with a woman that doesn't turn into a battleground."

"You have only had bad ones with a stepmother and a wife."

"Yeah, just the ones that count."

"Speaking of women, I guess you found something interesting to talk about with that pretty girl, last night? It sure took you long enough. Weren't you bone tired?"

Nolen smiled. "I'll admit she's attractive and has intriguing eyes. But don't try complicating my life right now. This is simply a business relationship. We just weighed the yellow." Nolen fished two plastic tubes of gold out of his sleeping bag. "You were snoring when I returned. Here's your share."

Digger hefted the tubes. "One of these is yours, boy."

"Nope. You and Hilda need the gold more than I do."

"I can't do that, son. Each tube must be worth about $4,500. That's too much for you to give away, considerin' the bills you have."

"Take the money; pay part of Hilda's hospital expenses. It's the least I can do. If you and Hilda hadn't told the social workers about what was happening, I would never have lived to see my eleventh birthday. And you didn't have to volunteer to be my foster folks either."

Digger remained quiet for several moments, as he stirred the fire. Nolen noticed him wipe his eyes. "Smoke blow in your face?" Nolen asked.

"Yeah, smoke." Digger grunted as he busied himself with his pipe. The old man began laughing.

"What's so funny?"

"You! Boy, your eyes were as big as saucers when you jumped through the pickup window yesterday. You looked like a big, wet frog trying to crawl into the cab. Pal was barking like mad. It's the funniest sight I've seen in years."

"You certainly weren't much help," Nolen replied. He watched the old prospector slowly stretch his arthritic back, then resume preparing breakfast beside a large redwood tree.

Digger was much like the spice scented redwood: old, tall, and wrinkled. He fit with the forest, sharing an inner peace with whomever was around. Nolen recalled other times the old man had joked with him, making sad moments disappear, turning his gloomy moods into something constructive. Now, Nolen decided, it was his turn.

"Digger, you really like this life. Even though you're always complaining about how tough the work is. Admit it. You enjoy going into the hills every month, driving through the prettiest parts of the Sierras, hunting for gold. You enjoy taking chances to make lots of money. And it keeps you healthy, hiking up and down the mountains while enjoying the peace and quiet."

"Peace and quiet? Rarely, with your nighttime yammering. But for over fifty years, I've never punched one of them time clocks. Not had to work for nobody. And I've sure avoided a whole passel of rush hour traffic."

"So, it has been a decent life after all?"

The old man sat quietly, thinking for several minutes, then slowly continued. "You're right about the chances. That's all it's ever been. Except once, a long time ago, when I found a specially good vein. That vein and the money to take better care of Hilda was all stolen from us."

"You never told me how you lost that strike."

"Letting your wife down is something a man don't talk about much. It was in Arizona. Two weeks after I filed on the vein, Hilda miscarried. Like always, we needed money. I got it by selling a portion of the profits from the site to a young lawyer I thought I could trust. I was so worried for Hilda that I didn't pay enough attention to the paperwork. When she came out of the hospital, I learned I had signed away the development rights. I got cheated out of that mine—legally!"

"That must have been tough for both of you."

"After Hilda got better, she needed to move on. She needed some new memories. So, I swore I would find her the mother lode in California. There, no one would take it away from us. Since then, I've studied real hard. I've learned how to protect a strike from them lawyers and con men. And along the way, I made sure you learned how."

"Digger, it seems you have been worrying about things from your past also. Is that why you've been driving us so hard lately?"

"Yep, I ain't got a lot of time now, to make good on my promise to Hilda."

5

Offer

The offshore wind swept across Monterey bay, whipping up white caps and crashing blue-green waves onto the rocky beach. Cool spray splattered Nolen as he passed the Coast Guard lighthouse, marking the ninth mile of his morning run. He increased his pace while enjoying the bright sunshine and the beauty of the jagged seashore below the flowered shore paths. He smiled, appreciating how his physical exertion had eased his tensions, allowing him to forget for awhile, his recent painful divorce. He sprinted the last quarter mile, discharging all his energy into the run.

Nolen cooled down by walking to Digger and Hilda's rust-stained mobile home. It was one in a line of 50-foot trailers, located three blocks from the shoreline. A sadness touched him, as he approached their lot. The small yard's normal tidiness was missing. Instead, leaves littered the tiny patch of synthetic grass which substituted for lawn. Crumpled, dirty pieces of newspaper bunched against the trailer foundation. Even Hilda's potted geraniums, which always outlined the yard with bright red blooms, drooped from an absence of her loving attention.

Nolen watered the flowers and cleaned the paper from the foundation. As he worked, he thought how difficult the last six months had been for his foster folks. Hilda had been in and out of the cancer ward several times. Digger was showing the strain of the long hours, spent vigilant at the hospital. Nolen's divorce proceedings had only contributed to their misery.

The computer search program was the only accomplishment preventing the situation from becoming worse. The probability maps had helped Digger and Nolen discover the placer gold in Maida's pool in January, and more at another site in May. Even though the gold petered out, it kept the old couple from applying for food stamps, while satisfying the doctors' demands.

Nolen stepped inside. "Digger, you home?"

"I'm in the bedroom. Be out in a minute."

The coal black dog rose from the kitchen rug and hurried forward, wagging his tail. Pal rubbed his body against Nolen's leg. "Digger, you know Hilda doesn't like Pal in the house." Nolen petted the canine, then pushed him outside.

"If you don't tell her, I won't," the old prospector said, when he emerged from the bedroom. He shuffled to the kitchen sink and attempted to fill a coffee pot with water. Finally, he lifted a stack of dirty plates out of his way.

"When's the last time you washed dishes?" Nolen asked.

"About four days ago. Why?" Digger grumbled.

"Well, it's an interesting way to decorate your counters, just before you go prospecting. It blends well with the stuff scattered on the floor."

"Don't be a smart-ass, boy. You know Hilda does the house work."

"Yeah, but it looks like I'll be stuck cleaning this place, before she returns home from the hospital next week."

"Humph, you might as well get used to it. You'll be doing your own house cleaning from now on."

Nolen stared through the kitchen window without replying.

"Son, I didn't mean..."

"Oh, it's OK. I know what you mean. In some ways, things will be much better. I'm just sorry my ex-wife chose this time to go back East to her folks. I have to ship the furniture, close bank accounts, and settle several other matters. I'd prefer going with you, as we had planned."

"While I'm away, why don't you visit Maida?"

"You really think she's something?"

"Boy, for a change, it wouldn't hurt you to be around somebody nice. She ain't afraid of hard work, spends wisely. Helpful little thing, too. Her advice sure saved us money on this year's taxes. And, every time we dredge the pool below her cabin, I notice you two get along real well."

"She's fun and has lots of sparkle. But I plan to avoid making the mistake I've seen many other divorced guys make—leaping into a relationship, before understanding why the last one failed. So, for now, I'll move slowly while appreciating some peace in my life."

"I said a visit—not start a relationship, as you call it."

Pal's barking from the porch steps caught Nolen's attention. He watched a shiny, black Mercedes Benz roll to a stop in the driveway. The 50-year-old driver straightened his tie, and checked the neatness of his razor-cut, gray-tinged hair.

"Digger, who do you know with a license plate titled, '24 Karat'?"

The old man squinted through the window. "He's some big shot. Works for CMR. Humph, he looks familiar. Might as well let him think I'm alone. Step into the bedroom, sonny, and stay out of sight. You might learn something."

Pal intensified his barking as the stranger emerged from his car.

Seconds later, the tall, slightly stoop-shouldered old prospector opened the trailer door. "Quiet, Pal."

The stranger spoke. "Mr. Vernon Johnston? I'm T.J. Tower, Vice-President in charge of exploration for

Continental Mining and Refining. I drove down from San Francisco today to speak with you. My assistant phoned you last week to confirm our appointment." Tower felt uncomfortable. Where have I met this old goat before? he wondered.

Digger and Tower surveyed one another. Tower's custom-made, three-piece suit contrasted with Digger's worn jeans and woolen shirt. The executive boasted a manicure; a stubble of gray beard marked Digger's weathered face. One man reflected affluence and the corporate world; the other denoted hard times and outdoor labor.

As Digger eyed the executive, an unusual toughness tightened the old miner's features. "Move, Pal," Digger said, shoving the dog's hindquarters with his foot. The animal slowly ambled off his customary step.

"The name's Digger Jorgenson, not Johnston. No one ever calls me Vernon."

A spasm of adrenaline raced around Tower's stomach when he heard the name. "Excuse me. I misplaced my note with your name on it."

"Come in."

Tower eyed the miner as they entered the trailer. After all these years, could this be the same guy? Tower scanned the clutter of knickknacks lining the window ledges, and wrinkled his nose at the sight of the messy kitchen.

"Want some coffee?" Digger asked.

"Sure. I would love a cup. Cream and sugar, please." Tower hated coffee, but he did not intend to refuse his host's hospitality. The old goat already seems angry, Tower thought. Hopefully, three decades, cosmetic surgery, and a name change will prevent him from recognizing me.

Tower fingered a small photograph, propped against the windowsill, next to the table where the executive sat. It showed a woman, a gangly boy, and a younger Digger, leaning against a battered pickup. To reduce the awkward feeling between them, Tower commented about the picture while Digger rattled around the kitchen. "Is this your family?"

"Yup, taken 'bout twenty-five years ago."

"How is Mrs. Jorgenson? I hear she is in the hospital, battling to overcome cancer."

Digger placed two mugs of steaming coffee on the table, sat down, and began packing his favorite pipe. "She's OK, now that the doc has removed a small tumor from her chest. Probably be home in a week or so. Your wife ever get sick and need special doctors?"

"Once, she hurt her knee skiing. Surgery was required."

"Bet you had all the money you needed to take care of her?"

Tower worried to himself, I must divert him from this subject, and then get the hell out of here—fast. "I didn't realize you had any children," Tower continued.

"We kind of adopted the boy just before that picture was taken."

"Does he live near you?"

"Gave up his career in the Army when Hilda's cancer got bad. Works here in town now. We always could depend on him to do the right thing."

"Digger, I don't like to beat around the bush. We want you to join us in exploring for gold in northern California. I'm aware you have hit two decent strikes within the last six months. That type of success is uncommon for a private prospector. However, I recognize your capabilities. You are the type of man we want for our exploration team. With proper financing and technical help from CMR, I'm confident you can bring in a large strike."

Digger inhaled from his pipe as he stared into Tower's slate-gray eyes. "I'm not sure I want any part of your big company."

"You would also benefit another way. I recently established a cash bonus program for prospectors affiliated with CMR. Depending upon the size of the strike, you would receive an immediate payment of up to $35,000, as well as a share of the claim. I'm offering you an opportunity to have CMR's support. We could help each other."

Tower tried, inconspicuously, to cover the gaudy, two-ounce gold nugget, dangling from a chain on his vest. Damn wrong day to wear the nugget that came from that Arizona claim, he worried.

"You know Whitey Jones?" Digger questioned.

"I have no idea who he is," Tower replied.

"I'm sure you don't. Hilda and I have been friends with Whitey and his family for years. Known them ever since they came to California, when the mine he worked in Montana closed. Way before you got here. For the last twelve years, he and his wife and his daughter have worked in the Newberg smelter. Now, they're all looking for jobs since CMR closed the refinery."

"Digger, those people feel bad about the smelter closing. So do I. In fact, I fought diligently to keep that plant open. You may not know that CMR maintained the smelter for five years, while we were losing considerable profit. We did that to care for the families depending upon CMR. Not many firms would be so considerate, Digger. And we are guaranteeing all our prior employees first priority for hiring when we re-open the improved plant."

"Whitey told me otherwise. He said CMR's never goin' to re-open that smelter. Instead, you're upgrading another plant in California, with newer technology. Also, your company didn't offer the Newberg workers any re-training, or even jobs elsewhere in the company. You just mailed them severance pay, and didn't even say good-by. Some of those men and women were within a year of retirement, too. Seems lately CMR uses people, just like any ore it takes from the earth. You extract all you can, then dump the worthless stuff on the tailings pile."

"Digger, I have made many of those same arguments. But you have been around this business for years, and must realize why these things happen. We experience only one good year out of several lean years. It also depends upon what the world market is paying for oil and minerals, which I can't control. Even though the company was forced to close the smelter, I'm sure you know, that over the years, CMR has been very fair with their employees."

"Yeah, I have to know what the large mining companies are doing. All wildcatters and independent prospectors watch the big firms. We track the technology you use, because we can't afford to develop it. And we try to avoid getting stepped on when two of you big guys get hungry for the same piece of dirt, and drive up the lease prices. Meanwhile, we stay alive spotting the jobs the large companies are unwilling to go after. The jobs with high risk or little chance for a large strike. Your educated engineers call them 'marginal return' projects."

Tower replied, "Everyone in this business is pressured to keep a constant eye on the bottom line."

"We independents know the big firms also keep an eye on what we're doing. You all recognize we drive harder, because it means food on our plates and clothes for our children. To survive, we try new ways and take chances. Many times, we help you big guys, by pointing where to go. I also know CMR hasn't hit anything new, of any size, in the last five years."

Tower reined in his irritation with the old miner's arrogance. "That is why I am here. You have the right skills. We can form a beneficial team."

"A lot of independents say lately CMR has been pushing them around. I've been told even the people in your own company disagree on how you do your job. No, I don't want to work with your firm. I don't want any claim I find, stolen."

Tower's anger flared at the blunt remarks delivered by the prospector. "You're right, Jorgenson. CMR has been suffering through some difficulties. And I have been forced to turn CMR around, during these last few months. But I'm in a far better position than you. My field men tell me you have used your share of the previous strikes to cover old bills and pay off past grubstakes. Also, this crummy little place has two mortgages against it, and you've borrowed the maximum you can from the banks. Yet, you still owe the doctors over $17,000 for your wife's illness. And she needs further medical care. You're in a tight spot, Jorgensen. Be reasonable! CMR is willing to make you a lucrative offer, in exchange for a share of any future discoveries. Plus, we'll lend you money and equipment to continue prospecting."

Digger glared at Tower. "Yes, I've got it tough—soul bending tough. But not so bad to sell out to a blood sucker like you, and give away any bonanza I strike. I won't let you use my skills, then cheat me like you did in Arizona!"

Tower's eyes widened. "I don't know what you are talking about."

"Your name was Tribou then. You were a fresh lawyer out of college, with a degree in geology, looking to break into big business. When we needed money, you claimed to be our friend, and said you would find a way to help."

"I have to leave." Tower stood and backed toward the door.

Digger jumped to his feet with a speed that surprised Tower, and grabbed the executive's tie. He yanked Tower's face close to his.

"No matter how you try to disguise yourself, with your fancy suits and titles, you always will be the worst kind of vulture. You live off wounded people."

"Let go of me!" Tower shouted, while attempting to pry Digger's iron grip from his clothes. "The mining company screwed you. I just helped get your financing. Anyway, it wasn't a valuable claim. The gold ran out after three years."

"What ain't valuable to a big company, could have damn well changed our lives." Digger slugged Tower, knocking him through the doorway onto the porch steps. As Tower scrambled onto his hands and knees, Digger kicked his buttocks, sending the executive rolling across the ground.

Tower rushed toward his car, then glared back at Digger, standing in the doorway. "Just like before, I'm going to clean you out! You are even weaker now. I don't care what it costs CMR. I'm going to block you from getting loans, or any other support you need to file another claim. I'll even pay doctors and hospitals not to accept your wife as a patient. She will pay for your mistake of rejecting my offer."

"I should shoot you for what you did to us!" Digger responded.

Tower jumped into his car. "I hope you hit it big, Digger. Because the larger it is, the more I will enjoy taking it away from you! Everyone will be applauding my success—while you remain in poverty. You'll come groveling to me, asking to be left alone. Then, I'll spit on you."

When Digger started down the porch steps, Tower stomped the gas pedal. He snatched his car telephone from its cradle and yelled into the receiver after his nephew answered. "Direct Special Projects to immediately begin following Digger Jorgenson. I want our men on this guy, like death on a corpse. Tell them I said to stay on this case, until I personally release them."

Nolen moved behind Digger and watched the Mercedes speed away. "Digger, are you OK?"

"I'm great. I've been waiting years to punch his ugly face."

"Yes, but now he's really dangerous, and he has the means and the manpower to make things very tough for us."

6

Accident

14 July

Desert dust blew over a Quonset-shaped bunker on an Army depot, located 200 miles northeast of San Francisco. What appeared to be a common ammunition storage area, hid one of only three biosafety level-five test chambers within the continental USA. Inside, a laboratory team leader made his last visual inspections of the three men standing in front of him. Each government technician stood in a bulky, white protective uniform. For easy identification, prominent, black, single-digit numbers were stenciled on their helmets, chests, and shoulders.

The team leader announced through his microphone, "Entrance procedures initiated at ten hundred hours. Begin communications check."

"This is Number 2, I hear you loud and clear."

"Number 3, commo is fine, boss."

"Number 4, every time we do this, I feel like a tightly wrapped mummy."

"Number 4, don't start your usual crap," the man in suit 1 replied. "This is serious."

"OK, Number 1. Let's call this episode, 'Return of Mummy Man'. Once again, he visits the tomb to terrorize little creatures."

The other two men chuckled.

"Inflate."

All four suits bulged outward from the internal pressure. The senior technician reached sideways and punched a wall button, causing a shower of disinfectant to douse his group. When the spray shut down, the men paired up, checking each other's suit. They searched for telltale bubbles that would reveal the existence of tiny leaks in the uniforms— leaks that would easily threaten their lives.

"Inflation status?" the group leader questioned.

"Number 2, OK."

"Number 3, OK."

"Mummy Man's sweating. This suit's like a hot house."

"Number 1, all OK. Pre-entry inspection completed."

While Number 1 moved to open a yellow door, leading into a control room, his teammates shuffled across the cement floor. From the center of the open bay, they retrieved three metal carts. Each custom constructed cart supported a cubical glass cage, confining an active brown and black monkey. An independent air supply automatically provided oxygen to the cage from a pressurized tank strapped to one of the cart's four supporting struts.

The furry animals scurried about as their prisons were rolled through the yellow doorway. They were pushed straight across the small inner control room area, and parked in front of another entrance. The second door was painted red and stenciled with a bold, white message: DANGER— ENTERING VIRUS LABORATORY. OBSERVER MUST BE PRESENT BEFORE ENTRY. OXYGEN MUST BE CARRIED BY PERSONNEL. — DANGER.

Number 1 waited until Number 3 locked the yellow door, which the team had just passed through. Then, he stepped to the right of the red entrance and depressed a button. A flash of ultraviolet light bathed the room beyond the red door with bluish-white illumination. Both Number 1 and Number 3 moved to a control panel positioned below a thermal-paned window, allowing an unobstructed view of the atmospheric self-contained room. They began preparations for the test.

Meanwhile, Number 2 and Number 4 unsealed the red door, and pushed the three carts into the center of the much smaller, innermost laboratory. Number 2 secured the red door from the inside, then Number 4 double checked the seal. They maneuvered the carts onto colored squares painted on the cement floor. Behind the colored squares, a wheeled pallet sat between two aluminum operating tables.

Inside the cages, the inquisitive animals watched the oddly clothed humans, walking about the antiseptic room. The men gingerly removed pairs of eighteen-inch, double-walled, silver canisters from a refrigerator. Each canister bore a yellow or orange painted band, which identified the type of gas they contained. A T-valve connected the tanks, facilitating the mixing of the two gases stored within the canisters. One monkey lost interest in the slow moving humans and snatched a piece of fruit from the floor. She began nibbling.

"This older one, on the end, looks like my mother-in-law," Number 4 quipped. "Eats like her too."

His partner stared at him. "You work at cracking jokes every time we run these verification tests. This stuff getting to you?"

"Nothing bothers 'Mummy Man'. Hand me that tubing."

Within minutes, Number 4 snapped the canister pairs into mounting brackets on the flat base of each cart. Then, he stretched a plastic hose from the canister valve, and attached it to a nozzle protruding under the floor of each cage. Next, they attached electrical power wires to the valves and methodically inspected their work. Finally, the two men left the innermost chamber and re-locked the door.

For the second time, Number 1 moved to the red entrance and stabbed the button. Startled by the bright light, the monkeys squealed and jumped about, chattering, then gradually relaxed.

"You know," Number 4 said to the team leader, "I could turn on the anti-viral light whenever I exit the test area. Save us all time, effort, and discomfort."

"I'm responsible for the safety of this team," the group leader responded. "Our safety procedures work, preventing a gas valve from turning on at the wrong time. Or, exposing us to any lingering toxic agents. I'm going to follow these rules to the letter, so I never have to explain any accident to your wife."

The senior technician returned to his console position and toggled switches, activating audio and video tape recorders. He verified that all three specimens clearly appeared on his video monitor, then spoke. "Beginning quality control test. Three randomly selected samples from..." He double-checked a piece of paper attached to his clipboard, stamped TOP SECRET. "Lot 2000-3 to be administered to rhesus monkeys. Verification of binary gas lethality is to be accomplished by visual observation, followed by an immediate autopsy. Sample-A was mixed 118 minutes ago. Sample-B was mixed fifty-eight minutes ago. Sample C is..." the technician flipped a switch on the panel, "mixed at 1020. The aged samples will be pumped into the laboratory in two minutes."

The other men in the room moved nearer to the window, as Number 1 studied the second hand of his stopwatch. "Administering samples now at 1022." After he turned three red console switches, a stillness seized the group. They did not move, merely listened and observed.

In the laboratory two of the monkeys began blinking their eyes and shaking their heads. Short, sharp spasms followed, jolting the two animals, as if they had received electric shocks.

The oldest female reached up to urgently scrub her nose with her forepaw. Suddenly, she slumped onto her side and lay panting. A moment later, she struggled to sit up, but her forearms would not support her weight.

The other affected monkey crouched on all four paws, leaning its head and shoulders against the wall of the container. Drool spilled from his mouth, and slipped down the glass wall of the cage.

The third animal seemed oblivious to the mixture softly blowing into its cage. It munched another carrot.

Gurgling gasps from the two infected monkeys filled each man's helmet. Number 4 pushed around the senior technician to flip the audio button. The laboratory sounds ceased, as Number 4 turned away from the window.

Nothing changed for several moments, then violent muscle contractions bent the two small animals at the waist. Brown streams of diarrhea laced with red blood shot out the anus of each of the two sick monkeys. From the safe side of the glass window, the men analyzed the noiseless dance of death.

Five minutes dragged by before the team leader turned off the red switches. "Secondary convulsions completed at 1027 for the test animals receiving the two and sixty-minute test mixtures. As designed, the 120-minute sample caused no reaction. We'll now conduct autopsies."

Following another flash of intense light, Number 2, 3, and 4 re-entered the inner test chamber, to begin their post-exposure examinations. Number 3 walked to the rear of the small room. He began removing scalpels and other surgical items from a drawer beneath the autopsy table. In precise sequence, he placed metal tools on both tables. Behind him, Number 2 and Number 4 disconnected the electrical power wires from the pump motors, manually verified all canister valves were closed, and then roughly yanked the plastic tubes from the metal connectors on the cage floors. Small puffs of dark gas spit from the tubes.

A super-sensitive molecular detector registered minute amounts of toxic agents present in the air. Immediately, red warning lights, hanging from the ceiling, began twirling while a wailing siren blared. The nerve jarring noise persisted, as Number 2 and Number 4 popped open the security clasps holding the rubber-edged, glass cage lids. They carried the lids and fitted seals to the aluminum tables.

Number 4 unlocked the wheels of the surviving monkey's cart. He rolled it to the side of the laboratory, beneath a TV camera. It would constantly monitor the animal during the next twenty-four hours to record if any incubation effects would appear.

The trio paused, when their team leader in the outer room ordered them to shield their eyes. He punched the button, igniting the brilliant ultraviolet light. With fitful coughs, the siren fell silent, and the red lights slowly stopped revolving.

Next, the men disconnected all the canisters, and carefully carried them to the pallet positioned between the tables. There they laid the paired cylinders side-by-side on the floor of the pallet, locking them within U-shaped fiberglass holding beds.

After Number 4 rolled the dead monkeys' cages next to the aluminum tables, Number 2 and Number 3 each removed a contorted body. They placed the animals face down, inside rubber body bags spread across each of their tables.

Number 2 and Number 3 began explaining their actions and what they subsequently saw. First, they used their scalpels to cut through the soft tissue at the base of the skull of their specimens. They expertly carved the sweaty skin away from the yellow-white bone surrounding the brain. Seconds later, the snapping whir of lightweight electric saws sounded, as the men cut through the protective bone. Wet, wrinkled tissue, inside the cranium, became visible. They peeled the bone from the brain, like the skin of an orange. Black and blue blood clots tattooed the outer surface of the organ.

Upon detecting the expected trauma signs, the men sliced down the side of each spine. As they wedged apart every third vertebrae, they spotted speckles of clotted blood along the spinal cord. Rolling the bodies over, they then attacked the chest cavity. They split open the rib cages to examine the lungs. Again, they found the telltale dark splotches. When they concluded, the two men utilized a special disinfectant to wash the blood and small pieces of skin from their hands. Then, they zipped the carcasses into the body bags.

Number 4 moved each bag to the pallet. He gently placed them between the gas canisters and two ten-gallon metal holding tanks. The tanks were connected by clear plastic hoses to the drain on the end of each operating table. Number 2 and Number 3 used the disinfectant to clean the surgical tools, then the cages, followed by the aluminum table surfaces. After expending several gallons of liquid to rinse their equipment, Number 2 and Number 3 signaled that they were finished. Number 4 unhooked the hoses, laid them beside the body bags, then wheeled the pallet next to the exit door.

With the experiment and autopsy completed, the men were anxious to move the excess liquid, bodies, and remaining gas to the incinerator bunker. There, they would destroy the toxic material. When the disposal was concluded, the team could remove their hot, stiff suits and relax.

The senior technician began talking his men through the slow tedious exit procedure. First, he verified that the primary gas sensor did not react. Then, he bathed the group in high-intensity light, followed by activating a second sensor.

It too, did not sound any alarm. With no evidence of any toxic agent, the senior technician authorized removal of the pallet from the laboratory. The team repeated the procedure in the outer lab. They moved into the preparation room, and again performed the triple-check. Finally, they slid open the outer door, permitting fresh air and sunlight to flood the mouth of the bunker.

Number 4 trudged uphill to a parked forklift, where he climbed into the driver's seat. For a moment, he enjoyed the serenity of the blue sky, dotted with clouds, and the open desert, overgrown with Manzanita. The natural comfort of birds playing among the bushes, and wind whispers rippling the patches of golden grass, did not help him quell his depression—a gloom induced by his participation in the death of another group of innocent animals.

He gunned the diesel engine to life. The box-shaped vehicle accelerated, rattling and bumping its way toward the pallet. The engine noise and movement prevented Number 4 from hearing the distant rumble, nor recognizing the growing, unnatural ground vibration.

Not until Number 2 and Number 3 began waving their arms and shouting, did he recognize the danger. He started to stomp the brake, but a sudden powerful earthquake jolt popped the forklift. His boot missed the pedal.

A noxious nausea gurgled in the bowels of Number 4, as he realized the forklift would slam into the pallet. He jerked the steering wheel sideways in a futile attempt to avoid disaster. His hope flickered as one of the forklift's iron claws missed the pallet. But the second one ripped through the line of binary canisters. Shafts of gas spurted into the air. They intermingled, creating an ugly, gray cloud, as the forklift crashed to a stop, pitching Number 4 forward. He bounced off the steering wheel, and then against the metal roll cage, ripping a gash in the shoulder of his protective suit.

"Help me! Don't let it touch me!" Number 4 screamed. He slapped his gloved hand over the rip and scrambled away from the darkening, deadly cloud. The bucking earth knocked him from his feet. On his hands and knees, he clawed his way upwind, where he lay praying.

When the earthquake's violent motion subsided, Number 2 rushed from the bunker to aid his panic-stricken partner. He yanked at a pocket flap on the pant leg of Number 4, to extract a spring-loaded injector. Without hesitating, he slammed the green tube against his friend's thigh, forcing antidote serum into the bloodstream.

"I'm going to die!" Number 4 wept.

"No, you won't," Number 2 replied. "Your pressurized air supply was pushing out of your suit, preventing any significant amount of gas from reaching your eyes or lungs. You were never in the main gas cloud. And the warning lights went out almost immediately, after this brisk wind moved the gas away from the bunker."

"One part per million will kill a man."

"You're protected. The antidote reached your system only seconds after the accident. At worst, you'll simply feel like shit."

The two men watched the toxic gas glide toward the desolate, unpopulated mountains, three miles north of the bunker. Serpentine tendrils snaked away from the ground-hugging cloud. Each changed color, first charcoal-black, then dark-purple, then fogging to misty light- purple. The singing of sparrows and thrushes abruptly stopped as the many birds flitting through the brush thudded onto the desert sand. Jack rabbits twitched among the rocks, as the expanding mist continued its uncontrolled drift.

"Thank God," Number 4 declared. "Everything is fine."

"Keep praying," Number 2 said. "Pray the gas dissipates before overtaking anyone downwind from us."

7

Promise

15 July

The jarring ring of the telephone sounded in Nolen's dark bedroom. "I promise to quit partying," he groaned. "Divorces suck." Shoving the confining covers aside, he tried rolling onto his stomach. A bulky form blocked his turn.

Who's there? he wondered. What was that gal's name? Someone's cousin from Montana. He slowly pulled the sheet off his head. When he dared peek, Pal's furry face stared back at him.

"Woof," the retriever barked.

"Get out of my bed, Pal!" Nolen pushed the resisting animal from the mattress and snatched up the phone. He ran his hand through his tangled, blond hair as he mumbled, "Hello."

The voice from the other end of the line sobbed, "Nolen. Nolen, this is Hilda."

Abruptly, he sat up. If Hilda was crying at 5:00 a.m., something was drastically wrong. "What's the problem?" he questioned.

"Nolen, I've got terrible news. Please come over—come quick!"

"I'll be right there." He jumped up, threw down the receiver, then flipped on the light switch. He yanked on his pants, slid his feet into his running shoes, and charged through the door.

Standing at the entrance to the mobile home, he paused to catch his breath from his three-block sprint. He feared bad news, as he stepped inside. I hope she hasn't had a relapse, so soon after being released from the hospital, he thought.

Blackness filled the trailer except for a small circle of light created by a wall lamp, illuminating the gray-haired woman. Hilda sat slumped in the kitchen chair, her rumpled house robe cinched about her. She leaned her elbow on the tiny breakfast table, supporting her bowed head with one hand. In her other hand, she clutched her rosary cross against her chest, as she softly prayed.

Nolen watched a tear roll down her wrinkled cheek, then fall through the silence and splatter on the table. Hilda looked at him with gaunt eyes. In her weak voice, she uttered, "Digger's hurt. He's hurt real bad!"

"How do you know that?"

"A nurse called, the Emergency Room nurse from the hospital in Susanville." Hilda hesitated, desperate to control her emotions. "She said Digger had an accident, a car accident. He's in...uh... uh...critical condition," she sobbed.

Nolen knelt and hugged the troubled woman. "Did the nurse say anything else?"

"He's had a heart attack. And he's only got a 30 percent chance of..."

Nolen cradled his foster mother. But his efforts were powerless to soothe her inner pain, or the wave of sobs which shook her frail body. "Don't worry, Hilda. We'll go to Susanville today. Digger will be OK. He's a tough old coot. Before long, he'll be here with you, cluttering up this place again."

"I can't believe it, Nolen. The day after you bring me home, Digger is struck down. We just begin to get back on our feet and then, more bills, more pain. I'm so tired of trying. How can God do this to one family?" She yanked her rosary from around her neck, splitting its beaded string, and threw the crucifix against the kitchen wall.

"I don't have an answer," Nolen quietly whispered, as he fought a growing tightness in his own chest. "I can only say, it's all right to feel angry and scared. Remember what you taught me years ago? 'Whenever times are tough, whenever we are scared or feel alone, realize God hasn't forsaken us.'" Nolen retrieved the cross, then pressed it into Hilda's hand.

She stared at the small cross she had carried since Digger gave it to her on their wedding day, almost forty-two years before. Over the decades, her callused hands had rubbed the cross shiny bright. During the troubled hours when her mother died, and at every Sunday mass, it had accompanied her. The three times she miscarried, and whenever Digger was ill, she had held the cross in prayer. Now, resentment waged a silent war inside her. She closed her hand around the crucifix, then leaned her head against Nolen's chest and cried. Even now, her hope survived.

He rocked the old woman in his arms. "You know I'll always be here to help. You can depend on me. I'll rent a private plane at the local airport so we can fly direct to Susanville. You must rest now, to regain your strength before we leave."

His foster mother continued to sob.

Five hours later, Hilda and Nolen hurried into the Susanville hospital. A bespectacled physician met them outside the Cardiac Care ward. "My name is Doctor Daly. I regret to tell you, it's unlikely Mr. Jorgenson will survive. He suffered a severe heart attack in the Emergency Room, shortly after he was brought in by the Highway Patrol. So, I'll let both of you speak with him, briefly. Try not to upset him. He's exhausted. A nurse will remain with you. She'll tell you when to leave. Agreed?"

"Yes, yes, of course," Hilda whispered.

Nolen supported Hilda as they entered the antiseptic room. Bandages covered portions of Digger's face and one of his arms. Tubes and wires extended from his gaunt form to ominous wall monitors. A splint protected his left leg. How old and tired Digger appears, Nolen thought. Even his skin has changed to an unnatural pallor.

Digger forced a faint smile when Hilda sat close beside him. With intense effort, he reached to touch her wrinkled cheek. "There's some things I must tell you. Precious, the way I'm feeling, I won't make it this time."

"Old man, don't you say such things. You'll be fine, just fine. I'll have you home for my chicken soup in a week." She squeezed his hand to assure him.

"Hilda, I love you. I'm sorry for all the hard times I put you through. All the worn dresses you had to wear. And the few cheap presents I brought home. I'm sorry...sorry for never finding enough gold, so I could buy you the things...to make your life easier."

"Digger, I don't look back and remember rough times. I remember shared dreams, working side-by-side. Silly man, I never needed fancy dresses to realize you cared. Your simple presents always made me laugh and then cry, because they proved you knew me so well. You always touched me gently and made me feel wanted. Those are the things that are important."

"It's not enough," Digger coughed.

"I'll never forget the night I told you I had cancer. We both cried for hours, you rocking me like a baby. The next morning, you picked me a handful of wildflowers. Remember? I dried those little flowers and put them in our bible to keep them safe. Every night, I touch them and thank the Lord for sending me such a good man."

Digger struggled to speak. "Hilda, there's another thing. I know in your heart you always felt it was your fault that we never had any kids. Don't...please don't feel that way. It was

God's choice to give us Nolen instead. I watched you share your love...to bring him out of his tight shell."

"It was you, Digger," she argued.

"No, not me. The boy was hurt by a woman. He needed to be healed by one. You taught him to believe...in himself, to trust, to care. The day he graduated from junior college you both made me so...proud of how...you made it happen. Even though you didn't birth him, you brought a good person to life in this hard world."

A grimace of pain contorted Digger's face. He motioned for Nolen to come closer. He whispered into Nolen's ear. "You must do something important. Almost found the big one this time. But...got sick before I could pinpoint the vein."

"Digger, gold is not important now. Quit chasing it."

"Nolen, take care of my Hilda. Promise you'll find it for her. She deserves it...I owe it to her. Taught you everything I know 'bout prospecting. You can do it. Look in the book...in the book. Promise me!" Digger insisted by tugging on Nolen's shirt.

"How did the accident occur? Was CMR involved?"

Digger squeezed Nolen's arm. "Give me your word."

"Yes, yes, I promise."

A racking cough seized the old man, causing the monitors to react with an insistent beeping. The nurse pushed the bedside call button, then placed her hand on Nolen's shoulder. "I'm sorry, sir. You must leave now. He's quite weak."

Hilda and Nolen left the room, as two doctors rushed to Digger's bed. A narrow window in the unit door allowed Nolen and Hilda to watch the medical team's hurried actions. Repeatedly, Hilda recited her prayers. A chill swept through Nolen when he saw the men and women slow their efforts, then stare at the monitors.

Moments later, Doctor Daly emerged. "Mrs. Jorgenson, your husband is still alive."

"Thank God!" Hilda exclaimed. "Oh, thank God."

"However, he has lapsed into a coma."

"When will he revive, Doctor?" Nolen asked.

"I can't say. Maybe an hour, maybe a year. It's out of our hands now."

16 July

By noon the next day, Nolen had finalized arrangements for Hilda to stay in a hotel, within walking distance of the hospital. She insisted on maintaining her vigil at Digger's bedside, until he recovered. She did not want him to awaken in an unfamiliar place, surrounded by strangers he did not know. And, Nolen also knew, she feared her life's love might die without his hand clasped in hers.

Reluctantly, Nolen began fulfilling his promise. He started by locating Digger's belongings at the police property yard on the outskirts of the small town. When he arrived, only one officer was on duty. The others were off for lunch, as Nolen had calculated. He did not want anyone watching him while he searched for Digger's notebook.

Through the years, Digger had learned that discovering a gold vein was more difficult than finding a wish in a wishing well. It required long hours, patience, and disciplined hunting. Digger always used one exploration technique to help him search systematically. He maintained a small ledger, detailing his efforts within a potential mineral region. The old miner's notes provided a valuable record, charting what areas of the land he had studied, as well as what signs of gold he had found.

There was danger in keeping a notebook after a discovery was made. Should someone steal the book, they could pinpoint the site. But Digger in his wisdom maintained several hiding places to protect his prospecting book from such a loss.

Nolen showed the Duty Sergeant his identification and the power of attorney signed by Hilda, authorizing him to retrieve Digger's truck and equipment. The sergeant glanced at the paperwork, as he selected a folder from a metal file cabinet. "All right, son. Here's a list of the property. Each item has been tagged. Check off the inventory, as I hand you the gear."

Within half an hour, Nolen had piled the items in the rear of a rented van. There, he rummaged through Digger's equipment cases. The notebook was not in any of the usual hiding compartments.

Nolen returned to the property room. "Officer, I want to check the damage to my father's pickup, to determine if there is any salvage value. Can you direct me to it?"

"Sure, just follow me." The policeman led Nolen to the back door, opened it, and pointed to one side of the fenced yard. "You'll find it parked in the far corner. Just bang on the door when you come back."

Nolen walked along several rows of dusty, damaged cars before he spotted the pickup. Three windows were shattered, the frame twisted, and the front, right side torn and crumpled. A lump grew in his throat when he approached the vehicle. He paused, blinked away tears, then pretended to inspect the exterior of the truck, while ensuring he was not being watched.

He crawled underneath. Caked mud covered the underside, forcing him to use his Swiss knife to scrape away the dirt. It took a few minutes to uncover a six-inch-square hinged plate and remove the three holding screws. He pried at the small bent door. It popped open, and a plastic bag dropped onto his chest. The valuable notebook and a plastic pint jar full of glittering gold lay inches in front of his eyes.

When Digger made his last request, Nolen believed his foster father was still chasing dreams. This quantity of shiny metal, however, made him question his doubts. His adrenaline began to pump.

Nolen thumbed through the notebook to the last few pages. The entries confirmed that Digger had arrived in the Susanville area five days earlier. He had been followed by some men, that he assumed were working for CMR. But Digger had been able to slip away to the prospecting area without being trailed. The entries for the last day were hardly legible. Some seemed to make no sense. Nolen concluded that Digger had narrowed his search after finding significant signs of a nearby vein. Then, the old prospector suddenly grew too ill to continue, so he decided to locate a doctor before searching further.

"How you doing there?"

Nolen jerked upright, slamming his forehead against the bottom of the jeep. He quickly stuffed the notebook and jar into his shirt and closed the little door.

The face of the sergeant appeared upside down next to Nolen's shoulder. "Sorry I startled you."

"It's OK. I just didn't hear you walk up." Nolen slid from under the vehicle and stood. "Anyway, I always needed a dent in my forehead, you know, like a dueling scar."

"The frame appears to be cracked," the officer said.

"Yeah, it's totaled. I'll send a tow truck over to haul it to a junk yard. Thanks again for your help." As Nolen hurried from the disposal yard, much was on his mind.

Fifteen miles west of Susanville, at the sprawling U.S. Army Western Ordinance Depot, a duty NCO stuck his head out of the communication room to call to the compound Security Officer. "Sir, you have a caller on the secure phone."

The black Military Intelligence officer hurried from his office. When he picked up the red telephone, he responded, "This is Major Clement."

"Major, state the college you attended, the last four digits of your social security number, and the number of children you have."

"Grambling, 1668, and no dependents."

"Roger. This is an oral message from Vinn Hill Headquarters, only for you and your commander. Make no written or electronic record of this message. Do you understand?"

"Yes, and I will comply with standard procedures."

"Message follows: Lieutenant Colonel Daly reported that patient Jorgenson went into a coma at 1045 hours yesterday. Daly handled the patient's relatives. They, and the Susanville hospital personnel, have no suspicions that Jorgenson was infected by a toxic agent. Daly assisted with the pathology, and ensured all documents indicate that the coma, fractures, and bruises were sustained in a car accident. Do you need me to repeat?"

"No, I understand your message." He hung up. "Sergeant."

"Yes, sir?"

"Record this call in your duty log as a routine monthly communication check. I'm headed for the Colonel's quarters."

At dusk, Nolen verified that Hilda had taken her sedative. Then he walked onto the rustic hotel's second-floor balcony. There, he selected a private seat away from the other guests. He studied the notebook entries again, and again, wondering what to do.

Should I tell Hilda that Digger may have found another strike? Or, should I avoid raising her hopes? This is the third time we've found gold using the probability maps. Crap, it's so unlikely! But, if we do hit a vein, Hilda could benefit from the best medical help.

Damn it, Digger, you knew I'd say yes. Were you asking only for yourself—a continuation of years of fruitless searching, not wanting to die feeling your life accomplished so little?

You and Hilda always expressed so much hope and determination. You both enjoyed the exhilaration of a promising find. And you both tried to hide your frustration when the small veins ran out. I refuse to subject Hilda to another emotional roller coaster. Digger, you did that to her countless times.

So, I won't say anything to her about the notebook or the gold. I'll sell the pint of gold to pay hospital and hotel expenses. Then, I'll prospect one last time, for you, Digger.

8

Search

By ten o'clock, the mid-morning heat cooked the highland desert, east of Susanville. Mirages danced in front of the jagged green mountains which outlined the horizon. The distant landscape appeared flat and repetitive. Nearby, the land's actual characteristics were apparent. It was yellow-brown in color, marred by many gullies and rocky ridges, spotted with yellow grass and hip-high brush.

An irritating, sweet, dusty odor scented the cab as Nolen steered his pickup toward the northern California hills. There, he hoped to locate the site where Digger last detected gold. The pollen triggered Nolen's hay fever. Repeatedly, he sneezed and rubbed his nose, as he compared Digger's far-off dream to the allergic reality of the moment.

The road curved eastward, skirting the Army depot, which spread across the breadth of the vast valley. Nolen glanced at the loyal retriever he had given his parents five years earlier. The animal sat next to him, stretching his head and shoulders through the window, sniffing the breeze. Nolen smiled, remembering how the furry puppy had cheered them. Now, Pal would be his companion, to guard his camp while he prospected. Nolen spoke to his pet. "If I were still in the service, we could save hours of travel time by cutting across the reservation, instead of going around. Well, Pal, it appears we'll reach camp late tonight."

The retriever chose to ignore Nolen and continued sniffing the wind.

Before dawn the next morning, the insistent, tinny rattle of the alarm clock roused Nolen. It had been too short a night's sleep. Nevertheless, now he needed to get moving. At sunrise, he planned to begin prospecting.

To test the morning temperature, he poked his nose outside the sleeping bag. The crisp sting of night air chased him back into the down-filled sack, where he enjoyed its seductive warmth.

Sleep captured Nolen, until Pal nudged him with his nose, then licked his face. Nolen recoiled from the surprise greeting. He pushed upright, so he could clutch his flannel shirt hanging on the side of the pickup. "All right. All right, I'm awake. Go away." The animal sat, wagging his tail, watching him.

After lighting a lantern, Nolen crawled from his sleeping bag to quickly finish dressing. He ignited a butane stove, then tossed a handful of link sausages into the frying pan, which sat ready on the grill.

The dog eyed the sizzling meat.

"Yes, Pal," Nolen said as he petted his friend, "some of these are for you."

"Woof."

The aroma from the cooking meat reminded Nolen of the many times he had prospected with his foster father. It was on just such a trip, he recalled, when Digger showed me how to pan gold. Now, I must employ every skill he taught me. From hilltops, I'll scan the search area to rapidly identify granite or quartz rock formations, where gold is often found. Next, I'll locate streambeds below the outcroppings. Then, I'll look for vegetation known to grow in soil containing gold. I'll mark all those spots on the terrain map, together with the high-probability sites, from my computer printout.

Finally, I'll drive across the territory, studying each site, to determine if they match any of the places Digger described in his notebook. I'll choose the best one, then dig there first.

Nolen reached for the frying pan. "Damn!" The hot handle seared his hand. He dropped the metal skillet.

Pal took advantage of the scattered sausages.

"Looks like you get all the meat, Pal."

The dog gingerly gulped the spilled links.

In preparation for the day's work, Nolen double-checked all the necessary items for the search—geological map, compass, food, water canteen, binoculars, gold panning tools, and electronic metal detector. Nolen scrutinized the most important item, Digger's notebook, before securing it in his coat pocket.

The paragraphs described several areas Digger had worked, and others that he had planned to investigate. Nolen assumed the gold that Digger had accumulated in the plastic jar was discovered on the last day he prospected. However, that day's rambling entry provided no clues to the exact site where Digger detected the precious metal, only that he was close to a vein.

The sun's bright crescent rose above the hills to the east, as Nolen rode his motorcycle to a nearby hill. There, he carefully scanned the surroundings with his binoculars.

It was a rocky, jagged landscape, mostly up and down. Northward, a 6,400-foot-high mountain peak dominated the horizon. Numerous ridges, resembling huge ribs, descended southward along the side of the hill mass. Sheer rock walls dropped to narrow streambeds, and dusty ravines twisted along the bases of the ridges. Thousands of boulders, split apart by the heat of the day and by the cold of the night, littered the ground. Spotted brown lizards and scrawny jack rabbits used the many crevices for shelter from the hungry hawks circling above. Weathered trees, scrub brush, and patches of grass clung to the earth in those spots where they could obtain sufficient moisture to survive.

Nolen penciled X's onto the geological map, corresponding to the descriptions in the notebook, regarding favorable prospecting sites. After his observations, he memorized a route to another hill a mile away, where he would conduct his second scan. Then he kicked the bike into gear and dropped over the lip of the ridge.

By two o'clock that afternoon, Nolen had finished his prioritizing. He decided to first inspect three streambeds below potential gold-bearing areas. If gold were in those places, traces would likely appear downstream from the source. The metal detector and panning tools would be most efficient there.

Nolen worked the first two creeks for four hours, without achieving positive results. He moved to the third, parking his motorcycle near a slow trickle, three-feet wide and a half-foot deep. There he swung his electronic metal detector back-and-forth to rapidly survey the gravel along the stream listening for any pinging that would signal a hot spot. At least here he found two hot spots that were more likely to contain gold. But Nolen controled his spirits knowing that often this method of sampling a site could provide false signals. He shoveled gravel into his green, plastic gold pan. Then, in the center of a small pool, he sat on a rock and began panning.

While keeping his arms and shoulders stiff, he used both hands to swirl the material in tight, counter-clockwise circles. The shaking motion would allow any heavy gold, present in slurry, to settle to the bottom of the pan.

The hard rock pained his buttocks. His back ached from the long day's labor. Nolen tried ignoring the discomfort, while shaking the pan and letting the mixture settle. He tossed aside a handful of worthless pebbles, poking out of the mixture. Again, he sloshed the pan back and forth in the pool. Each forward pass allowed muddy water to wash away. Nolen repeated the process, until only a tenth of the material remained. He leveled the pan and back-washed, by moving the water and remaining material in a clockwise motion. A thick layer of black sand coated the bottom of the pan. Nolen's lips tightened with disappointment, when no yellow emerged.

Smoothing the sand with one finger, he felt a pea-sized lump at the bottom of the pan. A ruddy, yellow nugget sparkled in the black dirt. Nolen's heart started beating rapidly. His fatigue disappeared.

His labors during the next hour produced ten more pebbles and an ounce of gold flakes. As the amount of gold grew, Nolen replaced his skepticism of Digger's last request with controlled anticipation. Perhaps there is a vein nearby? He did not have time to pursue an answer. Evening shadows darkened the creek. Nolen needed to take advantage of the remaining daylight to relocate his base camp upstream of the pool, to place his gear nearer to where he would be working the next morning.

At nightfall, he drove his truck into his new campsite. It was a grassy shelf, fifteen yards from where the small creek emerged from a narrow ravine, slicing through the hill. He backed his vehicle against a steep rock wall to gain a windbreak from the cool evening breeze.

Nolen began establishing his camp. First he fed Pal, then checked the engines of both vehicles. Unfolding the cot and sleeping bag, he began taking care of himself.

At the edge of the stream, he slumped to wash the dirt and sweat from his body. Frequent splashes of cold water helped keep him awake. Stumbling back to the truck, he opened a large can of beans and doused the greasy contents with hot sauce, his own special recipe. It did not help much.

He jerked awake when the half-eaten can of food tumbled from his hand, clattering across the stones near his feet. Nolen quit resisting, and slid into his sleeping bag.

At dawn, Nolen awoke stiff and sore, yet eager to begin. One hundred yards upstream from where he found the small nuggets the day before, he panned again. Gold appeared. He moved another hundred yards, and discovered even more yellow pebbles. After each successful panning, he continued upstream. During this process, he passed his campsite, and traveled one-half mile up the defile, until he no longer uncovered mineral traces. He had been following the gold trail to its source. And now, he just might have it bracketed between the last two panning points.

Stepping from the water, Nolen surveyed the creek area. The stream had sliced a winding, thirty-five-foot-deep, twelve-foot-wide ravine into the mountainside. The walls of the small gorge were composed of gray stone. Rocks and splintered branches littered the streambed. Cold water trickled around the boulders, periodically forming small pools.

Nolen edged downstream, looking for any sign of an ore outcropping. Halfway back to where he had last detected gold, he panned again. No yellow appeared. He continued his downstream search, while turning his head from side to side, inspecting the defile's features. Nolen's focus paused on a sliver of white quartz, barely visible, peaking from above a rock ledge, fifteen feet up the rock wall.

He climbed the vertical wall, using the few available cracks and jutting rocks for hand and foot holds. As his face rose above the ledge, he saw the two-foot-high, four-foot-wide seam of quartz encasing a mass of reddish-yellow material. "Oh, my God!"

Scrambling onto the narrow shelf, Nolen jerked a pick hammer from his hip bag and attacked the clear crystalline rock. Within minutes, he held a small piece of yellowish rock in his hand.

Nolen quickly verified the rock's hardness with the point of his knife. "Soft, and malleable," he nervously whispered. "Its color is consistent throughout. And the color doesn't change in sunlight or when shaded." His hands shook as he attempted to remove a clear plastic vial from his shirt pocket. It contained the colorless testing chemical, aqua regina. He dropped a tiny chip into the liquid. He held his breath. The gold dissolved.

Nolen jumped to his feet and started dancing in small circles. "I found it! I found it! Digger, I thought you sent me on another futile search. But, by God, I found it. Yahoo!"

His right foot settled upon several pieces of broken quartz. They rolled like marbles. Suddenly, he felt a terrible emptiness below his feet as his legs shot backwards. Fear churned in his stomach. He twisted toward the rock wall as his knees ripped against the lip of the ledge. Throwing his arms out, he bent forward. The ledge slammed his chest and face. Furiously, he scratched for a secure hold, while sliding downward. Jamming one hand into a small crack, he stopped his fall. Pain rippled through his entire body, as he dangled from the shelf.

He struggled farther onto the outcropping and lay quiet, letting his mind stop racing. On his back, he rested, trying to catch his breath, allowing the adrenaline to wash out of his system. Really smart, he chastised himself. That stunt nearly cost you your life! Lucky to have survived with only minor cuts and bruises. Should have known better.

Digger had related hundreds of stories to Nolen about prospectors who hit large strikes. But then, for strange reasons, they lost their fortunes. Some failed to accurately identify their mine location. Others were injured returning to civilization. Some were murdered. Many had their claims stolen. I swear, he thought, none of that will happen to me!

I'll use my many years of training to safeguard this discovery. I won't fail Digger and Hilda.

Carefully, Nolen climbed up the remaining twenty-foot portion of rock wall to the top of the defile. There, he used a Global Positioning System receiver to verify the location of the strike. He repeated the procedure, until he was satisfied that he had not made any mistake. Then he entered a description of the quartz seam and its map coordinates into Digger's notebook.

Although it was painful, Nolen carried additional equipment from the campsite to the top of the defile. He knew making a detailed vein search would be difficult, but it was vital for determining the claim boundaries. Manzanita bushes covered the hillside above the ledge. The intertwining branches of scrub brush forced him to employ a machete to clear the necessary pathways across and along the ridge.

He knelt in the center opening of a rectangular, six-foot-long earth-penetrating radar and slipped the carrying straps over his shoulders. He struggled to his feet with the awkward burden of heavy batteries and electronic boxes. Excitedly, he began tramping down the narrow lanes, interpreting the pinging signals, while attempting to maintain his balance.

Crisscrossing the hillside for hours, he concluded the vein stretched one-half mile east of the creek, and one-quarter mile to the west. His crude measurements indicated that it varied in size along its length. In several places, it appeared to be several feet wide. In other spots, it narrowed to a few inches. The seam was massive.

He continued working, hiding a three-foot high stone marker in the brush thicket above the vein outcropping. Under the base of the marker, he buried a small can. The can protected a paper notice listing the date, his name, and site identification information. After bounding the claim with twelve other camouflaged rock pyramids, he returned to the ledge.

Ignoring his aching arm, he sank three pitons, establishing a secure anchor for a climbing rope. The rope would allow him safe movement from the streambed to the ledge, as he removed gold ore from the head of the vein.

During the balance of the afternoon, Nolen chipped quartz away from the gold vein. As dusk fell, he trudged into camp with forty pounds of ruby-yellow rock packed into his rucksack. After dumping the load into his pickup, he massaged his cramped back and neck, hoping to relieve the throbbing tension in his muscles. He winced when one of the many blisters on his fingers burst.

Plopping onto the grass, Nolen petted his dog while appreciating his recent labor. "Pal, this definitely ranks as the third best day of my life. Rescuing a helicopter load of orphans from the Iraqi Army is still number one. That sensuous summer evening when I first made love to a woman remains number two. Finding this gold vein, definitely, qualifies as number three. And if it contains high-grade ore, it'll be the largest gold strike ever made in the United States. Hard to believe, Pal. Now, all I have to do is prevent it from being stolen, before I file the claim for Digger and Hilda."

9

Night Flight

14 August

An Army reconnaissance airplane sped through the ebony night. The bulbous plastic canopy and radar equipment, slung under its belly, gave the aircraft a bug-like appearance. Patches of clouds whipped along the sides of the plane. Inside, green light from video displays highlighted two soldiers.

The pilot spoke to his equipment operator, "We'll cross Checkpoint Tiger at exactly 0206. Are you ready?"

"Roger that. I have a GO for all radars, thermal scanner, radios, tape recorder, and map readout equipment." He paused, waiting for the search area to appear on his displays, then said, "Checkpoint Tiger is now on my screen. Let's do it."

Fifteen minutes into the mission, the observer broke the silence inside the cockpit. "I just recorded a thermal flash, one mile west of our flight vector. Vicinity coordinates 454648."

"What is it?" the pilot asked.

"Several heat sources. One, probably man-made. Fly to the southwest, back across the ridges, and I'll scan it again."

Within minutes, the aircraft shifted course to a new search pattern over the rugged, mountainous region. "Bingo!" the observer yelled. "Thermal target at coordinates 455650. One vehicle...one person...a campfire, and possibly a small child or animal. I have double confirmation from both the thermal and radar screens. Target is within security zone Whiskey."

The observer toggled several switches, as he spoke into his helmet microphone. "Apple one, this is Panther three seven, message follows. Are you prepared to receive high-speed copy?"

The reply came, "Panther three seven, this is Apple one. Send it." With a push of a button, the digitized sighting data descended to the ground station within two seconds.

Major Clement rubbed his chin, as he stood in the middle of the communication center, considering the implications of the information just received from the surveillance aircraft. The black officer had executed his duties at the depot for several years. Never before, during a special operation, had someone been spotted so near the military reservation, in a position which would allow him to observe their nighttime activity. Major Clement worried about this unlikely coincidence. One thought comforted him, as he looked around the room. The long console, crammed with sophisticated perimeter-monitoring equipment, was manned by his dedicated and well-trained men.

Fortunately, two years earlier, higher headquarters had accepted his recommendation to use National Guard aircraft. The airborne surveillance greatly augmented the fences, guards, seismic sensors, and thermal optical viewers employed to prevent entry into the heart of the depot. Legally and unobtrusively extending their security zone had paid off tonight, providing early warning of a potential intruder, squatting in civilian territory. Major Clement read the message several times and contemplated contingencies. Minutes later, he issued a warning order to his two-man security team to prepare to depart on patrol.

The Major then boarded his vehicle, seeking the detachment commander. The camouflaged, cross-country truck wound its way through the long rows of ammunition bunkers. Each storage bay was a reinforced concrete box, covered with compact soil which created a tapered, humpback shape. Each measured sixty-foot-long, forty-foot-wide, by twenty-foot-high. Every bunker had only a single sliding steel entrance, intended for trucks or forklifts to enter to load their dangerous cargo. Strict materiel-handling procedures were followed to ensure that, should an explosion occur, only one bunker, and those few men working in it, would be lost.

Over 500 bunkers spread across the dry sandy depot. All had identical internal cavities, except six used for special operations. The jeep rolled to a stop adjacent to one of these special bunkers.

A red light bulb, barely interrupting the darkness, protruded above the steel personnel entrance. When the security officer hopped from his vehicle, a soldier stepped through the faint cone of light. The soldier did not aim the M16 rifle at the Major, but held it in position to be quickly raised and fired, if necessary. "Halt! State your name and purpose."

"Major Clement. I'm here to talk with the Colonel."

The soldier whispered the evening challenge, "Lincoln?"

"Duck," replied the officer.

"Correct, sir. You can enter now." The MP opened the door.

The security officer strode down a short hallway to a well-lit, solid metal entrance. He slid a plastic identification card through the scanning device. After a green light appeared above a bank of buttons, he punched in his code number. The door slid away, revealing a wide, sloping corridor.

At the end of the corridor, he peered through a window, viewing the loading activity which was occurring further inside the underground chamber. Using the intercom system at the window, Major Clement contacted his commander within the pressurized portion of the interconnected bunker complex. "Colonel, can you come outside for a few minutes? I need to talk to you, sir."

"Major, I'm trying to finish this inspection, and supervise the loading of these trucks. I don't have time to do an exit procedure. What do you want?"

"Sir, our air surveillance has reported a campsite within security zone Whiskey. I believe we should scrub tonight's mission."

"What? Why?"

"For three reasons, sir. First, Iraq and Kazakhstan have recently increased satellite coverage of this region. Second, a hostile source may have learned something about the accident during the last mission. And third, they could be using the individual at the campsite to provide confirming visual intelligence. Therefore, sir, I feel we should discontinue the transfer while an unknown person is compromising the security zone."

The Colonel considered the situation. "Major, if we stop this shipment, we will miss two deliveries—back to back. That will put the supply program three months behind schedule. Since the daily gaps in satellite surveillance now only provide two half-hour windows for nighttime pickup, I don't want to cancel this consignment. Anyway, we have no hard evidence that any hostile nation is aware of this operation. Send a patrol to recon the camp. Let me know if anything develops."

"Sir, I strongly recommend we abort this mission."

"Major, I understand your concern. One thing you didn't mention is that the Muslim satellites may soon go to 100 percent coverage of this area and drastically slow our efforts. Now go carry out my order." The Colonel resumed inspecting the trucks parked inside the bunker.

Major Clement's jaw muscles tightened as he complied with his instructions, while mentally dissenting. Damn 'Can do' attitude. The Colonel is determined to get back on schedule, since the accident. But he's bypassing the wise precautions which have kept this project secret for five years. Shit, I hope he's right. We sure don't need to blow it now.

A subdued rumble woke Nolen, as he lay in his sleeping bag. "Am I dreaming? What's making that noise?" He pushed himself to a sitting position, then shook himself to ensure that he was awake.

A jet transport flashed overhead and dropped below the ridge south of his camp. "Hell, that plane is flying far too low!" He listened for an imminent crash, but he heard only the soft night wind. Nolen dressed quickly and hopped onto his motorcycle, then raced up the side of the ridge. A steep rock wall blocked his approach to the top. He switched off his headlight and parked his bike, grabbed his binoculars, and scrambled up and over the obstacle.

From the highest part of the ridge, he surveyed the valley below, expecting to view a flaming wreck, but saw only darkness. Then he heard the faint sound of a diesel engine to his left. Scrutinizing farther to the east, he waited while his eyes adjusted to the darkness. Gradually, he distinguished the outline of a large plane with several trucks parked around it. People were hurriedly loading crates into the jet's faintly-illuminated cargo bay.

This is strange, he thought. If that plane hadn't flown right over my camp, I wouldn't have heard its descent, since its engines sounded muffled. That airstrip inside the military reservation doesn't appear on my maps either. Any landing lights or tower are also camouflaged. Maybe, it's just a training exercise? Yet, it seems more like some of the clandestine missions I was involved with overseas. Well, it's none of my business. And, as long as they stay out of my hair, I'll stay out of theirs. With his curiosity somewhat satisfied, Nolen returned to camp.

Twenty-six minutes later, the jet departed from the desert airstrip. And three more times during the night, the reconnaissance airplane applied its scanning sensors to confirm that the intruder still remained within the security zone.

14 August

As sunlight kissed the mountain tops, a man dressed in a drab-green Forest Service uniform crawled forward. He progressed along the ridge south of Nolen's camp, until he reached a pile of gray boulders. Wisps of cotton-white, morning ground fog rose into the air. He used his binoculars to study the stream below, as well as the man moving around the campfire. Using a hand-held radio, the observer contacted his partner. "Haul up the gear," he ordered.

Ten minutes later, another man crawled toward the rock pile. On his back, he lugged an uncomfortable rucksack. As he slid forward, a reel spilled telephone wire behind him. Reaching the rocks, he removed the pack, pulled a foot of cord from the reel and connected it to a re-transmission device, packed within a side pocket of the rucksack. After the telltale beep verified the secure radio was working, in the jeep hidden at the base of the hill, he transfered the microphone to his partner.

The observer spoke into the hand mike. "Apple one, this is Spring Bird nine. Spot report, over."

"Spring Bird nine, this is Apple one. Send your copy, over."

"This is nine. We are in position, 400 meters southwest of the camp. It's occupied by one Caucasian male, about five-feet-eight-inches tall, medium build, light colored hair. He's wearing a holstered pistol. Present activity, he is eating breakfast. Two vehicles are parked in the camp. The first vehicle is a dark-gray Ford pickup with a rear camper shell, and a motorcycle mount welded to the front grill. I'm unable to see any license plates from this vantage point. The second vehicle is an off-road motorcycle, red and white, with leather saddlebags. Appears to be a 250cc size. A large black retriever is also at the site. I can't tell if it's sentry trained. One sleeping bag and other camping gear are visible. Over."

"This is one. Anything else? Over."

"This is nine. I don't know if the target planned it, but it is very difficult to observe his camp. It took us three tries before we could safely watch him. We'll stay here, unless the target breaks camp."

"This is one. Get the license plate number ASAP."

"This is nine. Wilco, out."

The two men maintained their vigil from their high-ground location for several hours, then radioed another report. "This is nine. The target has established a routine of leaving for approximately 40 minutes, then returning with a rucksack loaded with material, which he dumps into the rear of his truck. We're going to relocate our observation point into the creek, downstream from his bivouac site, to obtain better intelligence. We can't get too close because he leaves the dog in camp while he's gone, over."

"This is one, roger, out."

When Nolen again left his camp, the two men executed their plan. Thirty minutes later, they positioned their jeep behind a clump of pine trees off the side of the creek. Then they hiked upstream another quarter mile to a bend some 300 feet away from the camp. There, the observers waited.

"The wind is blowing toward us, and that dog hasn't alerted," the junior sergeant noted. "Even if he is sentry trained, he can't smell our scent."

"I'm going to transmit the truck's license plate number," the first observer answered. "Are you ready to take photographs? This guy should show up in about ten minutes."

"I'm set. With my telephoto lens, I should be able to count the number of whiskers growing from his chin. I'll also take a picture of whatever he's collecting in the back of that pickup."

10

Race

A slight mist permeated the air as Nolen lowered his heavy backpack filled with gold, to the base of the rock wall. This is my sixth load for the day, he thought, as he wiped his brow. That's enough. I'll be lucky to ever stand upright again. At camp, I'll wash this smell off, then drive to a restaurant to celebrate. A cold beer and a hot meal will be my reward.

But first, I must close and camouflage the gash in the rock wall. I don't want anyone stumbling onto this vein. Not before I file Digger's claim.

Nolen scooped broken pieces of quartz and loose dirt back into the hole. On top of the sandy material, he packed several large stones. After climbing down to the streambed, he swept fallen quartz flakes into the stream. Only then did he feel comfortable, that the mouth of the mine and all signs of his diggings were adequately concealed.

He sat down next to his backpack and hunched it onto his blistered shoulders. Rolling onto his hands and knees, he struggled to his feet to begin his downstream trek. Digging tools jangled together in a leather bag banging against his hip, and his coiled climbing rope hugged the top of the rucksack. Combining the equipment with the gold made the load his heaviest of the day. One more time to build character, he rationalized.

Even though Nolen had a well-designed backpack, the shoulder straps painfully dug into his sore muscles. Within minutes, rivulets of sweat streaked his body, as he sloshed through the water and soft sand. Several hundred feet later, he cursed. "Damn, it's hard this time."

A distant clap of thunder caught his attention, enticing him to study the dark clouds washing over the mountain peak. What a way for this day to end, he thought. Those rain clouds are blowing in my direction and dumping buckets of water. I'll be drenched and chilled by the time I reach the truck.

He hastened his pace, then stopped dead still looking toward his feet, recognizing that the depth of water in the creek had risen to his calf muscles. He jerked his head around to stare upstream. Adrenaline slammed into his system, as he detected an approaching low, steady rumble, rushing down the ravine.

Nolen slapped the rope off his pack, yanked the bag of tools from his hip, then bolted downstream. His thinking shifted into high gear. The world around him began moving in slow motion. He no longer felt the sweat beading down his chest, nor the ache in his shoulders. The loose footing under his boots and the boulders he bashed his legs against were meaningless. Only two things registered in his mind: his slow, piston-like leg movements and the growing, frightening crescendo of destruction chasing him.

He rounded the last bend in the creek before his campsite, understanding the danger of his defiant choice. "I'm not dropping Hilda's gold!" he yelled.

The howling of the flash flood chased Nolen, as he struggled up the rough bank and lunged toward his truck. The hurricane, preceding the wall of water, blasted sticks, dust, and pebbles off the ground and hurled them downstream like thousands of machine gun bullets. Tons of churning liquid followed the fury. A muddy-brown wave exploded from the mouth of the gorge, smashing everything within its path, gouging a new streambed, spinning rocks and boulders as if they were lightweight toys.

Nolen saw none of the sight. Fear clamped his eyes shut, while his strong arms locked around the front bumper of the pickup. Churning water raced up to his back, sucking him toward the center of the deadly torrent. It shook him back and forth like a limp rag doll. He clung to the truck and prayed.

After what seemed an eternity, the roar subsided. The deadly tug of the water lessened, then disappeared. The feeling of slow motion departed. Nolen relaxed, exhausted but thankful for the blessed silence.

Panting and spitting sand from his mouth, he rolled over, releasing the heavy rucksack. He felt totally battered. His jeans were shredded below the knees. Both boots were missing. When he stood, he winced, as a shaft of pain signaled that he had strained a back muscle.

He surveyed his new world. The front of the truck now angled down, two feet lower than just minutes before. Luckily, the motorcycle had been parked farther from the creek, preventing it from being swept away by the wild current. But Pal was nowhere in sight.

"Here, Pal! Come here!" Nolen shouted, worried.

The retriever's bark echoed from a distance. Minutes passed before Nolen spotted the dog, slowly limping toward him. He ran to hug the animal, then checked for injuries. Pal's hindquarters were soaked, and he yelped when Nolen squeezed his right rear leg.

"It's a good thing I didn't set up camp closer to the creek. Otherwise, I would be mincemeat now. Looks like we both reached high ground, just in time, Pal. We're just bruised. We'll be as good as new in a couple of days." Nolen scratched Pal's ears.

The dog wagged his tail and licked Nolen's face.

"Pal, the next time I work the vein, I'll pay closer attention to what is happening around me. This is the second time I almost killed myself, and you too. Let's return to civilization—some place safer."

Several wheel spinning attempts finally freed the pickup from the sandy hole where it had settled. Nolen then slung the motorcycle onto the rack, mounted on the front of his truck. After eliminating the few remaining signs of his campsite, he drove away. As the vehicle bumped along, he whistled a happy tune, grateful just to be alive. He ignored the pain in his back.

The flash flood had re-sculptured the creek. Banks, boulders, logs, and sand, all had been repositioned. Nolen maneuvered his vehicle around the piles of debris, careful to avoid the soft spots in the newly carved creekbed. Rounding a sharp bend in the stream, he came upon a nearly nude body, wedged upside down, between a log and a large pile of rocks.

Nolen jumped from the truck to examine the ripped and shredded corpse. The unnatural limpness of the body revealed most of the bones were broken. The only identification appearing on the dead man was a U.S. Forest Service shoulder patch, dangling from a jacket sleeve still on the cadaver.

Nolen began searching for additional flood survivors. As he slogged downstream, he called out, then listened for any response.

Just as he ceased his search, he heard a faint moan drift from underneath a brush pile beside the stream. Removing splintered logs and broken tree limbs, Nolen discovered a second man, lying face down. A large boulder covered his hips. His blood stained his pant legs, as well as the surrounding sand.

Slowly, the injured man turned his pale face toward Nolen. "Help me," he gasped, clamping his hand around Nolen's wrist. "Please, help me."

"First I must roll this rock off you. We must stop the bleeding. Then I'll take you to a hospital."

"Use...use the radio, in the jeep. Call for a chopper," the man pleaded.

Nolen looked left, then right. "Where's your jeep?"

"At the bank. Under some brush..." The man's words were choked off, as a death rattle slithered from his throat.

Nolen peeled the dead man's fingers from around his wrist. He walked downstream to the next bend. There he found the vehicle spilled on its side, with its green and white Forest Service logo partially sanded away. He pried open the bent passenger door, seeking the radios. Inside, the two transmitters were cabled to a KY-42-Alpha secure message encryptor. "Why the hell are Forest Service guys using secret military communication equipment?"

Nolen raced back upstream, returning to the last dead man. He shoved the boulder aside, and swept sand off of the man's hips. From the pants pocket of the corpse, Nolen pulled a soggy wallet. In it, he found a Forest Service identification card, as well as a green military ID, designating his rank as Sergeant First Class.

"Sarge, I don't know why you were using the Forest Service for a cover. But, I'm not sticking around to find out. I sure don't need the military interfering with my site filing." Nolen shoved the wallet back into the man's pants. Minutes later, he gunned the pickup out of the creek and began racing down a dirt trail, determined that no one would prevent his departure.

Major Clement waited for the detachment commander to take his seat at the conference table, then began his report. "Sir, both sergeants are dead. Their bodies were retrieved forty minutes ago. They were thrown about during a flash flood. At least, it appears that's what happened. Tire and foot prints, found in the creek, indicate the security zone intruder found both men, as well as their vehicle. Sand scraped from Thurman's body revealed that the suspect looked in our man's wallet."

"And, why is that important?" the Colonel questioned.

"Thurman had his military ID in his wallet. The suspect may have seen the identification card and realized our men were not forest rangers, sir."

"Son of a bitch," the Colonel swore, simultaneously slapping the table. Major Clement braced himself.

The commander jabbed his finger at the security officer. "Damn it, Major. That was your team out there, and you are responsible for it. Such an error should never have happened!" he shouted.

The Colonel sat silently for several seconds, attempting to control his anger. Then, in almost a whisper, he continued speaking. "Major, I'm well aware of the excellent work you have accomplished during the last several years. I recognize this is the only mistake your team has ever made. But it's a damn, serious mistake. Let anything like this happen again—and you'll be counting snowflakes in Alaska for the balance of your career."

"Yes sir! I clearly understand."

"Anything else?"

"Sir, our helicopter recon hasn't located the suspect, only vehicle tracks heading out of the stream. I recommend we find the individual to clarify whether or not he is a foreign agent."

"Just how do you propose we do that?"

"Sir presently, I have the helicopter checking the main roads in and about Susanville. Additionally, I have ground teams searching for the intruder or his vehicle in the nearby towns. If successful, they will shadow him until police support arrives. Meanwhile, we can request that our Washington D.C. contact implement counter-intelligence plan 'Quick Shield'. We can further provide our contact with the suspect's vehicle license plate number which Thurman reported prior to the flash flood. Then the police, together with the FBI, can track him down. Meanwhile, we complete our mission."

A resigned look crossed the commander's face. "Murphy's law is in full effect. First, an earthquake strikes while a fork lift is moving toxic material, causing us to bust open a canister and leak gas all over Hell and back. And now, I have a possible spy!"

"Well Sir, one good aspect of this situation is that it's happening when our operation is almost finished."

"Perhaps you're right. All we have to do is hold the lid on this baby for another three or four months. Then, it will be a great military success. OK, get me a typed synopsis of the facts concerning the suspect. Let me know immediately if our teams locate him. Meanwhile, I'll notify the FBI to help find and detain our mystery man."

11

Assay

16 August

Nolen spent the night hidden in the center of a recreational vehicle camp, surrounded by travel trailers. His excitement to learn the value of his discovery forced him awake at 4:00 a.m. Impatiently, throughout the early morning, he marked time, watching laughing families emerge to prepare breakfast under the tall pine trees and enjoy the nearby cascading mountain stream.

Finally, he left the camp, descending along a winding, two-lane road. It provided a clear view of Susanville, clustered on the western edge of the expansive highland desert. Reaching the town, Nolen parked in front of the only commercial assay and testing laboratory in the county. He waited for the lab to open, drumming his fingers on the steering wheel, observing the town come to life. When the doors were unlocked, he hurried into the office.

At the forms table, he completed an assay request, then moved to the service counter for assistance. Shifting his weight from foot to foot, he waited anxiously for the clerk, who was standing at the office coffee machine, studying the morning newspaper. Nolen called to the chemist, "Excuse me, can I trouble you to run an assay?"

The man gave him a resigned look, rested his cup down on his desk, and approached the window. "You must leave your sample, and come back tomorrow," he stated bluntly.

"Well, you see, I'm in a hurry, and I certainly would appreciate getting the results this morning. I'll gladly pay extra for the service." Nolen offered the assay request, along with two twenty-dollar bills, to the chemist.

The man reviewed the form, nonchalantly slipping the money into his pants pocket. "Let me see the sample." Nolen relinquished several chips of ore, taken from across the face of the gold vein.

The chemist ripped a carbon copy from the analysis form and handed it to Nolen. "Going to be awhile. Better take a seat."

The chemist ambled around a laboratory table, casually gathering the material and chemicals to perform the required series of tests. Impressed with the results, he raised his eyebrows, then strolled into a rear office. Seeking a private conversation, he closed the door behind him, then made a telephone call.

When he returned to the laboratory area, the chemist finished typing his report. He dropped one copy of the document into his desk drawer, then spoke. "Mr. Martin, I've completed the analysis."

Nolen's hands trembled as he accepted the paperwork. He scanned the report in order to understand the composition of the sample. His eyes widened as he read:

Gold	84.5%
Silver	9.0%
Iron	5.0%
Quartz	1.1%
Trace minerals	0.4%
Total	**100.0%**

He marveled at the typewritten words, turned away from the window, and wandered toward the door. How incredible, he thought. Digger was right. The mother lode does have an eastern branch. After chasing his dream all his life, Digger deserves to show this to Hilda! Nolen stopped, turned back, and said, "Uh...I need the sample."

"Here you are." The chemist handed Nolen a small plastic bag containing the gold. Nolen moved away, but again returned to the counter. "You sure it's point 845 pure?"

"Sure, I'm sure. I ran the assay twice. Don't often see many samples that good. Did you discover it around here?"

"Ah...yeah," Nolen stammered. "Yeah, I found it...west of here in...in a creek...alongside the road. Uh, excuse me, I've got to go."

Two men entered the assay office as Nolen rushed out the door. The tallest one sprouted a bristly, black Manchu mustache and wore a battered, straw cowboy hat. His shorter companion was burly, bearded and bald.

The clerk acknowledged the two men with a nod. The smaller man followed Nolen outside and watched him clamber into his truck. The bald man walked past the rear of the pickup, studying its interior cargo. A green blanket covered a lumpy pile near the rear door. Yellow sand and pebbles lay scattered along the edges of the cloth. The man continued across the street and scrambled into a car. When Nolen drove away, the man merged his vehicle into the local traffic.

Inside the assay office, the big man settled into a chair. As he propped his boots on the clerk's desk, he gruffly inquired, "What'd you learn about this guy?"

"First, give me my $1,000," the clerk insisted.

"For what? I ain't got any real information yet."

"Look, Luke, you're trailing that young fellow. I already gave you his name and home address. Over the phone, I told you that his sample is almost eighty-five percent pure gold. That's plenty of important information. It's well worth my price."

"Here's $250. Before you get any more cash, I want more meaty specifics."

"You Continental mining shits are all tight-fisted bastards!"

"Stop the compliments, and give me some real information."

"Well, the guy damn near pushed me over, when I unlocked the front door earlier this morning. For assay, he gave me an eight chip sample, totaling 2.24 ounces. Best high-grade stuff I've ever seen. The kid later tried to mislead me about where he found the sample. Wanted me to believe it came from a creek west of here. But I can tell the difference."

"How so?" Luke asked, with obvious interest.

"In the first place, if the sample had been placer gold, it would have been smoothed and rounded, after tumbling along the bottom of a creek. The chips he gave me were ragged-edged, hard rock gold. In the second place, the sample contained trace salts, found only east of Susanville. Now hand over my money. That's all the information I've got. Which, I might add, should be enough for a bloodhound like you."

Luke laid down another $250. "Give me a copy of that report. Then you get your last $500."

Nolen followed a side street, as he searched for a newspaper stand. "OK, Pal, let's find out how much our pile of gold is really worth." The retriever snapped his jaws at an annoying fly, buzzing around the cab.

"We need three values: the purity of the gold, the number of troy ounces of gold we have, and the world spot price presently being paid per troy ounce. Now, the assayer just gave us the purity figure. The second figure is easy to calculate, since there are 14.58 troy ounces per pound. Hm-m-m, I'd say, I mined 415 pounds of gold, giving us, let's say, 6,050 troy ounces. So, if I can find a *Wall Street Journal,* or some other newspaper publishing the spot metal prices, we'll have the third figure. Pretty slick, huh, Pal?"

"Woof," Pal retorted, seemingly responding to his master's enthusiasm.

Nolen parked the truck beside the entrance to a liquor store, jumped out, and purchased a newspaper. He tore through the business section, until he located the gold prices quoted for the previous day. The metal was selling for $502 per troy ounce. He tossed the paper toward a nearby trash barrel and rushed to his truck. On the back of the assay report he wrote: .845 x 6,050 x $502. He punched the numbers into his hand calculator, paused a second to calm himself, then hit the equals button—$2,566,349.

"Yahoo!" he hollered, waving his arms in the air. Nolen wrapped his arms around his dog and kissed him on the muzzle. Pal growled at his owner.

"Next week, Hilda will see the best cancer specialist in California. And you, my mangy friend, will have steak to eat any day you wish. But first, we must hide our gold, and start preparing for the gold rush!"

From across the street, the CMR observer witnessed Nolen's excitement. When Nolen left the parking lot, his follower unhooked a CB microphone. He informed his partner to quickly travel south, out of town, along the state highway to catch them. The bald man kept glancing into his rear view mirror, until he spotted his companion approaching. Moments later, Luke's pickup sped around the lead car, and he assumed the tailing effort.

The CMR agents halted their vehicles at a bend in the road, when Nolen wheeled into a remote gas station. The smaller man left his car and walked forward to the pickup. "What's the story on this guy, Luke?"

"His name's Martin. Connected with Jorgenson. The one the San Francisco office alerted us to watch for. You know, the old man we were following. The one who had the car accident."

"Didn't we drop that case when Jorgenson went into a coma?"

"Special Projects ordered me to stay on it, until some big shot says to stop. Someone's real interested in what Jorgenson was after. Maybe, this guy really has found something?"

"What did he have assayed?"

"Lode gold. Eighty-five percent pure."

"Jesus H. Christ! I looked inside his truck, just as I left the lab. He's got a blanket covering a large pile of stuff. Yellow sand and chunks of quartz were scattered all over the floor. Shit, I bet there's several hundred pounds of ore under that blanket."

"Let's grab him and see," Luke urged.

"What do ya' think Special Projects will say?"

"Screw them. We do all their dirty work, and all they ever pay us is peanuts. This time, we're following some guy who has a fortune stashed inside his truck. More money than you or I ever saw. So, we're going to take that gold for ourselves. And, Special Projects can kiss my hairy ass!"

The bald man snatched a pair of binoculars from the seat of Luke's truck and scanned the station. "He went into the restroom. Looks like that kid pumping gas, is the only other person around This may be our best chance. We'll nab him while he's in the shitter."

Nolen was zipping his fly and shouldering open the restroom door when Luke's fist suddenly smashed into the side of his face. The unexpected blow spun Nolen back into the urinal.

The two CMR men jumped on their victim, tied his hands behind his back, and jerked him upright. To gain more space within the narrow room, they shoved him onto the toilet seat.

Nolen attempted to revive from the blow, as the big man rifled through Nolen's jacket pockets. "Shorty, you were right. Look at this!" Luke held up a lump of gold the size of his fist.

The bald partner grabbed Nolen's pickup keys. "I'll check his truck." He hurried to the front of the gas station, where the teenage boy was washing the windshield of Nolen's truck. Walking to the rear of the truck, the CMR man unlocked the tailgate window.

Pal sat on the passenger's seat sternly eyeing the stranger. When the man reached inside the pickup, Pal lunged into the back of the vehicle snarling at the intruder. Startled, the CMR man jerked away, instinctively slamming the door shut.

The dog's sudden, insistent barking caused the curious attendant to step to the rear of the truck. Seeing the stranger, he yelled, "Hey fella, what are you doing? This ain't your truck!"

The short man took several rapid strides forward and struck the boy twice. The youth's knees buckled. He bounced when he hit the cement, then lay motionless. The CMR agent dragged the teenager into the garage. He hid the boy from view in a small tool room. When the bald man emerged from the maintenance bay, a station wagon, driven by a mother with two small children, stopped beside the row of gas pumps.

In the restroom, Nolen licked blood from his swollen lower lip. Ignore the pain, he told himself, through a semi-conscious fog. He shook his throbbing head. Words, drilled into him years before by a military instructor, popped into his mind. "The best time to escape is immediately after capture."

Nolen moaned and slumped backward, as the man rummaged through his pockets. Though his shoulder sockets burned with pain, Nolen lowered his wrists downward into the cold toilet water. He needed to wet the rope, so it might stretch or slip apart enough for him to free his hands.

The big man jerked Nolen upright, shoving a knife hard against his face. "Sit up, pisshead, or I'll cut you real bad!"

Anger filled Nolen, yet he remained still, except to cautiously twist his bleeding wrists, attempting to loosen the bindings without being noticed. His attacker began thumbing through Nolen's wallet, tossing papers and pictures onto the floor. A few minutes later, Luke opened the restroom door to peer outside.

Nolen did not hesitate. He stood and whirled toward the man. As he spun, he brought his right leg parallel to the ground. Luke turned back toward the toilet, just as Nolen snapped his leg forward with all his strength. He slammed the heel of his boot into his assailant's chest. A sharp crack pierced the air, as a rib broke. A desperate gasp followed.

The big man doubled over in pain, dropping the knife and the wallet. Nolen regained his balance and struck again. He used his powerful leg muscles to drive his knee upward, against the man's face. Blood sprayed from Luke's nose as his body straightened from the force of the blow. Nolen kicked once more, crashing his foot into the man's groin, knocking him to the floor.

Nolen leaned his ear against the door. He listened to hear if the other man was returning. No footsteps approached. Frantically, Nolen tugged at his bindings. The rope, holding his hands together, loosened. However, he could only separate his wrists by a few inches. He squatted. When his hands were positioned below his butt, he then plopped onto Luke's chest. The big man moaned; he was beginning to regain consciousness.

Nolen slipped his hands toward his feet, but could not clear the rope over the heels of his boots. His frustrated yanking merely tightened the coils around his wrists. Sweat dripped into his eyes. Re-assessing his situation, he deliberately calmed himself, then began pushing one boot off by using the toe of his other boot.

The bald CMR agent stood by the gas pumps, swearing under his breath, as the station wagon stopped for service. The female driver rolled down her window while her son and daughter pushed and pulled each other, in playful frenzy, across the rear seat. The agent forced a smile and said, "What can I do for you?"

"Fill the tank with premium, please."

"Yes ma'm; I'll have you traveling down the road in a few moments."

"Mommy," the young boy whined, "I need to go to the bathroom."

The agent spit. "Lady, my toilets aren't working. They backed up this morning 'cause a rat drowned in one. It's a real mess in there right now."

Grimacing, the woman ordered her son to wait.

The CMR man finished pumping gas and accepted the woman's cash. Relieved, he watched her drive away. After checking that no other cars were coming, he hastened into the garage. There, he found the remnant of an old greasy blanket, which he wrapped around his left arm. He seized a rusty piece of pipe from a junk-filled cardboard box, then approached the driver's side of Nolen's pickup.

Pal jumped onto the passenger seat. He curled his lips back from his teeth and snarled. The man positioned himself, keeping his hip against the driver's door, in case the dog lunged. Using his right hand, he unlatched and pulled the pickup door slightly ajar. Then the agent jerked the door open, and thrust his left arm toward the waiting animal.

Pal leaped forward without hesitation, sinking his teeth into the blanket, shaking the covered arm in violent motion from side to side. In response, the man dragged Pal into the door opening. Viciously, he clubbed Pal over the head with the pipe. The animal's front legs collapsed. His jaws released their grip. The agent relentlessly hammered, again and again, until blood drained from Pal's battered head. When the animal's body twitched, the agent struck one more time. He watched the canine slip onto the floor of the truck, before he slammed the door shut.

Satisfied that the dog was no longer a threat, the bald man rushed to the rear of the pickup and opened the tailgate. He flipped the blanket from the pile. In full view, the richness of the ore made his mouth gape open. Quickly, he re-covered the gold. After glancing up and down the road to discern if his criminal activity had been observed, he rushed toward the restroom.

Nolen was using Luke's knife to cut the wrist bindings, when the metal door swung open. The CMR agent's eyes widened, surprised to see Nolen sitting squarely on Luke's chest, almost free. He hesitated, only momentarily, before lunging forward, slashing the pipe at Nolen's head.

Nolen ducked left, under the bloody club. In defense, he drove the knife into his attacker's stomach, then ripped the blade upward. Intestines gushed over his hand. Blood flooded onto the floor, as the man dropped the pipe and fell forward. Nolen crawled from under the squirming, dying body and leaned against the wall. Anger faded from his consciousness. Hell, he thought, the feeling after killing another man never changes, simultaneous relief and repulsion.

Nolen retrieved the gold chunk, his car keys, personal papers, and wallet out of the coagulating pool of blood. He tore off his gore stained socks, and shoved his bare feet back into his boots. Using his shirttail, he wiped his finger prints from the knife. After wrapping the big man's hand around the hilt of the knife, he tossed it behind the commode. Then he confiscated the two men's wallets, thrusting them into his coat pocket. Finally, he washed away most of the blood which had spilled across his arms and legs. When he finished, he stepped near the door, listened, then looked outside. He was relieved to see no witnesses to the noisy incident.

Nolen ran to his truck. Just feet from the driver's door, he noticed the dark puddle of blood, spreading across the cement. "Pal. Oh my God, Pal!" he yelled. "Please, God let him be alive." A sick feeling overtook Nolen, as he felt the chest of his battered pet. There was no heartbeat. His friend and loyal companion, his Pal, was gone.

He screamed to vent his anger, "You lousy bastard!" Nolen whirled and stomped toward the restroom. Struggling to gain control of his grief, he stopped.

What about the boy? Did they kill him too? Nolen ran into the garage to search for the attendant. He found the unconscious teenager. The youth moaned as he tried to raise his head.

"Kid, stay here till the cops come. Stay here. Understand?" The boy slumped back to the floor, as Nolen stepped to the office telephone. He reported a knife fight at the gas station to the operator, hung up, wiped his fingerprints from the phone, then ran to his truck.

He gently pushed his pet aside, and snatched a road map off the dashboard. To get away—to protect the gold, he knew he must travel fast. However, he needed to avoid the main highways. That would reduce the chance of being stopped by the state police, if they started searching for his truck. After identifying a series of secondary roads leading toward San Francisco, he threw down the map, and wheeled the truck onto the roadway.

Pain crawled through Luke's body as he regained consciousness. He felt a gooey wetness, but he could not identify it, until he opened his eyes and saw blood and organs steaming on his legs. His partner's grotesque face stared at him. Sickened, Luke kicked the body away. He wanted to vomit.

A sharp pain lanced across his chest, making him gasp, pause, and hold his ribs, until the hurt subsided. Damn, everything's screwed up, he thought. I'm in a real jam. I need help. Maybe, I can get CMR to believe Martin murdered Shorty? I'll tell them Martin panicked, when we asked him about the gold in his truck. No, that's not a good enough story. Hell, I'll worry about that later. Now, I've got to get out of here.

Despite the severe pain in his chest, Luke staggered through the doorway and stumbled toward his truck.

12

Transformation

Nolen stood beside the fresh dirt mound, not wanting to leave. A claw of pain gripped his heart. I'm so sorry Pal, he silently grieved. Ever since the day I carried you home as a pup, you were my special friend. Hilda and Digger always appreciated your affection, and will miss your loyalty. Nolen blamed himself for Pal's cruel death.

He wiped his moist eyes while walking back to the truck. He had hidden it in a grove of trees, abutting a logging trail. Nolen recalled the day's events, trying to determine how the two men could have learned of his gold. Hell, I wasn't around anyone, except in the assay office. Those two guys who jumped me must have been the ones I passed, as I left with the analysis report. I was so excited, I wasn't paying much attention. Who were they?

Nolen leafed through the contents of the two wallets. A faint carbon paper entry on a motel receipt caught his eye. He stared intently at the paper. Above the 'representing' line, were three letters—CMR! He looked up, shaking his head. Of course, CMR. Now, it all began coming together.

How could I have been so dumb? That chemist must have alerted the CMR heavies, while he prepared my assay. Digger was right, Continental and Tower are blood-thirsty vultures. They swoop down to steal whatever they want. If

Tower is responsible for Digger's condition, I'll make it my life's goal to see justice served.

But I can't do anything about that now. I must leave this area, and definitely avoid the police. They might arrest me for the homicide at the gas station. Or maybe haul me in for questioning, regarding the death of those soldiers near the military reservation. Being involved in two separate incidents where people die would certainly be difficult to explain.

The cops would be skeptical over my claims of innocence, even though the big man's fingerprints will be found on the knife. And I couldn't prove I didn't steal the gold, without revealing the location of the gold vein. Word of a massive strike would leak out before I could file my claim. Thousands of people would flood the Susanville region. CMR might even find the gold mine. Hell, I don't need to be this popular!

Over the years, Digger taught me how to safely bring in a mine. But that was fantasy. Now, I'm dealing with reality. Unexpected twists and turns are occurring faster than I ever envisioned possible. Nolen wiped perspiration from his upper lip.

If I'm going to fulfill my promise to Digger, I must change my behavior. I can't keep viewing my world as secure and civilized. It is not! Making foolish mistakes only endangers me and those around me. Instead, I need to constantly realize I am hunted, anticipate danger, and instantly react when attacked.

A shudder shook Nolen. Damn! I have to turn loose the iceberg side of my soul. That part of me I continually work so hard to control.

Closing his eyes, he cleared his mind of all present day thoughts, by evoking the memory of a savage childhood experience. Nolen began reviving old hurts and buried emotions...

It was a warm evening when his parents delivered him to Juvenile Hall, society's words used to describe a jail for children and teens. His stepmother jerked him from the car by his arm and dragged him up the front steps. As unconcerned as if she were discarding a sack of worthless garbage, she shoved him toward a short, fat matron. Feeling lost and afraid, Nolen struggled to prevent any tears from falling, as he watched his mom and dad drive away. They never glanced back.

Nolen looked up at the matron, who was scrutinizing him. He felt like a bug on a slide. He followed her into the unfamiliar building. At a steel desk, she stubbed cold her cigarette and asked his name. He squeezed it out, trying to hide his fear of the unknown. She wrote his name in a ledger, adding him to the bottom of a long list of names. Within seconds, he had joined a unique group. He had become one of her social charges.

The matron led him through the building. During the night, half the lights were turned off, leaving the building in partial darkness. They moved in muted silence, from dark to light, to dark again, their shoes sliding across the marred linoleum, emitting an occasional squeak. The only other sounds were the repeated rubbings of the woman's flabby, nyloned legs, and the monotonous jingle of the keys she carried.

It was not his usual world, but an alien one, with fences and walls and people bounding his universe. Distance took on a new meaning for Nolen. Now, it was not how far he could throw a ball, nor how far he could run, but rather how far would he be allowed to go. He felt his heart pulsing.

They stepped into the mouth of a lighted hallway. Three solid steel doors were embedded within one wall. They were painted chocolate-brown, six and one-half feet high, three and one-half feet wide. Rivet heads tattooed the surface of the metal barriers. A narrow slide was built into each door, so any activity by the boys inside could be monitored. Like a demon opening its jaws, one portal swung wide, while the fat woman smiled, patted him on his shoulder, and mouthed words he never heard.

Slowly, he stumbled into the cell. His eyes darted, surveying, wondering. An older teenager occupied the chamber. A chill crept up Nolen's spine, as he realized the woman was closing the door. There was a swish, followed by metal crashing against metal. Seconds later, the lights went dark. It was insanity. It was reality. He was caged.

Nolen clambered into the empty top bunk bed and curled against the corner wall, hugging the pillow tightly against his face to muffle his sobs. His tears were lessening when he felt a hand sliding along his leg, and heard a voice saying he would feel better in a few moments.

Nolen shouted at the larger teenager, warning him to stay away. Lashing out with his small fists, he angrily jabbed his molester's face. His cellmate smiled, an evil taunting grin, while urging him not to resist. "The night supervisor has left for the other side of the dormitory. She can't hear your calls for help." The older boy came forward again. Nolen began yelling, brawling, and kicking with all his pent-up fury.

Minutes later, the matron, who had lingered in one of the nearby hallways, heard the ruckus, and returned to separate the boys. She dragged her new arrival to another cell...

As the frightening memories faded, Nolen chanted three phrases, over and over.

"No one loves me."

"Think and stay calm."

"Fight to survive!"

The technique had aided him several times through the years. Gradually, a chilling, dead feeling re-surfaced. He was transformed. His features had not changed. Now, they were merely a facade, masking the suspicious animal dwelling within.

"T.J., there's been a positive development in the Jorgenson case," Tower's nephew remarked, as he entered the executive's expansive office.

Suddenly interested, Tower looked up from the memorandum he was reading. "Yes?"

"Our Reno office just reported that Nolen Martin killed one CMR agent, and broke the rib of another."

"Who is the survivor?"

"Luke Daniels. This Luke fellow says the incident occurred when they were talking to Martin at a gas station, near Susanville. They were following a lead provided by a local assayist, who notified them Martin had brought in a gold sample. Luckily, Daniels fled from the gas station before the police arrived. He feels confident he won't be linked to our other agent's death."

"Wait a minute," Tower said. "A guy does not kill someone without strong motivation."

"Well, Daniels stated that the assayed material contained eighty-five percent gold and nine percent silver. Our man also viewed several hundred pounds of ore in the rear of Martin's truck."

"All right!" Tower shouted. "This is the break I've been waiting for. Old Jorgenson's boy obviously found plenty of high-grade gold, somewhere around Susanville. Soon, I'm going to own a valuable gold mine."

"Controlling the mine will guarantee that the Hokido Board of Directors selects you as the next company president."

"I'll have power, estates, and beautiful women, all the priorities of life."

"T.J., what about the dead agent?"

"Oh yes. Hum...we will have to conceal any CMR involvement in this murder. We need someone with the stomach for tidying up messy situations." Tower rubbed his chin, considering who might fulfill his needs. He snapped his fingers as a thought came to him. "Call my cousin Boodan Tribou. He is usually camped in the oil fields in Louisiana or Texas. Our Houston office will help you find him. Provide Boodan with all the facts concerning the Jorgenson case. Explain that I want him in California, to locate Martin. Tell the greasy Cajun to see me the moment he arrives in town."

"I'll make the call right away."

Tower reclined in his chair, grinning with confidence. All I have to do is capture Martin. It will be a pleasure extracting the location of his strike. What a satisfying revenge that will be against the old miner. Jorgenson will regret ever insulting me.

Five officers and three civilian scientists gathered around a rectangular table in the Commander's conference room. Major Clement handed each man a manila folder. He moved behind a podium at the front of the room, waiting for the Colonel to finish paging through the update report. Silence filled the chamber, a stillness highlighted by the sound of a wasp banging against a windowpane, struggling to attain his freedom.

The Colonel massaged his right shoulder. The recent rainy weather bothered the collarbone he smashed in a helicopter crash twenty years earlier. He nodded.

Major Clement began briefing. "Gentlemen, we supplied our Washington D.C. contact with the vehicle license plate number, suspect description, fingerprints taken from the ID card, and all other available data. They returned the following information:

"Source 1: California Department of Motor Vehicles. A Photostat copy of the suspect's driver's license is provided in your folder. His name is Nolen Martin, SSN 532-41-0038, age 35. A Caucasian, height five-feet-eight inches. Weight 180-pounds, blond hair, green eyes. He resides in Monterey,

California, a coastal town, situated about 100 miles south of San Francisco."

"Now we know the identity of our mystery man," the Colonel interjected.

"Source 2: Army military records. Martin enlisted in the infantry in 1984. He sequentially graduated from airborne, ranger, explosive ordinance demolition, and anti-terrorist schools. He received a direct commission to First Lieutenant in 1991, the result of his performance in Iraq during the Desert Storm war. He had fifteen years honorable service, but resigned his commission as a Major nine months ago."

The Colonel slumped in his chair. "Are you telling me that one of the Army's best-trained soldiers just happened to be outside the post during our special operations?"

"Yes, sir."

"It can't be mere coincidence," the Colonel complained. "Could he have joined the CIA, after he left the Army? Tell me more about his terrorist training."

"As a Sergeant, Martin was assigned to the Blue Light counter-terrorist group at Fort Bragg, North Carolina. Records indicate he was adept at his work. For example, in 1989, he individually detected and dismantled a terrorist bomb, smuggled into the world economic summit, held in Venice, Italy. He saved the lives of two presidents, six prime ministers, and one king. After the summit, he was promoted to Staff Sergeant—ahead of his contemporaries.

During his service, he was awarded the Royal Hero's medal from Kuwait, a silver star with V device for valor, meritorious service medal, purple heart, three Army commendation medals, and a letter of reprimand from his Division commander."

"Tell me about the Kuwait medal and the reprimand," the Colonel directed. "That's an unusual combination for any soldier."

"Sir, during Desert Storm, Martin was part of a team sent into Iraq, the night before hostilities began. The team's mission was to destroy communication links to the underground command and control center, operated by the General in charge of the Republican Guard Tank Corps. To preclude having their communication antennas attacked by aircraft, the Iraquis encircled the equipment with children taken from Kuwait City.

"Martin's task, during the American raid, was to move the innocent nationals away from the antennas, before the towers were destroyed. He accomplished his assignment, but he would not leave the kids behind. He detained one of the extraction helicopters on the ground, until he loaded two women, eight children, and most of his squad. The bird was so heavily loaded, it had difficulty lifting off in the thin desert air. Martin and two other sergeants remained behind. They exposed themselves to withering fire to divert Iraqi attention away from the fleeing aircraft. Martin was wounded in the arm. He and his men escaped into the desert, where they were later rescued. Martin's actions jeopardized the success of the mission."

"And that's why he got the letter of reprimand?" the Colonel asked with a puzzled expression.

"Yes, sir. But, when the helicopter landed, the team commander learned they had retrieved the six-year-old grandson of the King of Kuwait. The Iraquis had threatened to kill all members of the royal family that they held, if the US-Arab coalition attacked. The Kuwait royal family was so grateful, they contacted President Bush, thanking the U.S. The King specifically commended Martin and the other Sergeants for their gallantry. A month later, Martin was promoted to First Lieutenant."

"Major, what type of person is this guy?" the Colonel wondered.

"Sir, I have included copies of his efficiency reports, security background investigations, and his last military photo. He is a dedicated, self-motivated individual, exhibiting strong self-confidence, unafraid to challenge senior officers when he believes he is correct. As well as being technically proficient in military skills, Martin's superiors state that during dangerous or crisis situations, he thinks extremely well on his feet."

"He certainly demonstrated those traits in Kuwait," the Colonel agreed.

Another officer, seated at the table, spoke. "This guy had a good shot at making general. And he left before he earned his full retirement pay. Such action is strange. What explanation did he provide in his request for retirement?"

"He stated that an illness in his family forced him to return to civilian status."

"Major, have you been able to learn what he has been doing as a civilian?" the Colonel questioned.

"Sir, our third source, IRS records, indicate Martin has been employed by a computer firm. Credit investigations reveal he has no apparent major debts."

"How did he get involved with computers?" the Colonel probed.

"He had two years of college, before he joined the Army. After he received his commission, he completed his college degree by attending night school."

"Damn, this guy has talents that bother me! Go on."

"Source 4: California state police. At 1305 hours today an all points bulletin was issued for a late-model, gray Ford pickup with a front-mounted motorcycle and California license plates."

The Colonel frowned. "Sounds like our suspect's vehicle. Why the APB?"

"Sir, a man was found murdered at a gas station, eleven miles west of Susanville. Someone had sliced open his stomach with a knife. The attendant claimed that a man fit-ting Martin's description was in the station just prior to the killing."

A scientist, seated at the table, spoke. "It appears a third party is now involved. Any clues as to who that might be?"

"Some," the security officer responded. "The murdered man's car was found, parked near the gas station. During the last year, he has been employed by Continental Mining and Refining. CMR is a large mining company owned by a Japa-nese conglomerate."

"Major, break in any time with the good news," the Colo-nel ordered, in a sarcastic tone.

"Well...ah...sir, there really isn't any."

"Yeah, yeah. Go on with your report."

"Gentlemen, our aerial surveillance, our ground patrol reports, and these latest sources of data, allow us to infer two conclusions...neither of which are good news.

"First, Nolen Martin has peripheral knowledge of our na-tional security operation. This conclusion is based on several facts. He was observed inside security zone Whiskey. Our transport jet passed directly over his camp. He handled the disguised NCO's identification card. Footprints at the jeep prove that he saw the secure radio equipment. Combining this information with his extensive military experience should allow him to easily deduce something unusual is transpiring at the Depot."

"What we can't determine," the Colonel added, "is whether or not he would deliver such information to the press, the Japanese or some other group."

Major Clement continued. "The second conclusion is, that we might soon encounter teams of CMR personnel, dig-ging for minerals near our reservation boundary. This nega-tive scenario is a possibility, because we know Martin extracted some unknown material from the creek. Then he later was involved with, and perhaps killed, a CMR employee."

Another scientist inquired, "What is the chance that Martin found something of value in the creek?"

"We have no evidence that he did," Major Clement replied. "Our team scoured the stream area, specifically looking for signs of digging. But the flash flood had washed away any indications, if indeed, they ever existed. Yet, this complication of the dead CMR employee tends to indicate that he may have found something of value."

"However," an officer at the table interrupted, "Martin's activities at the creek may have just been a cover activity. Possibly, he was attempting to determine what cargo we periodically fly out of here. He might be working for someone who could jeopardize the secrecy of this project."

The room fell silent for several seconds, before the commander spoke. "Let's see if I can simplify this, Major. There is a guy running around who is smart enough to blow the whistle on us—someone involved with an unknown third party. And I do not know where he is, or what the hell his intentions are?"

"Correct, sir."

"Shit!" the commander exclaimed. "Any recommendations?"

"Yes sir. First, we initiate 24-hour production, in order to complete our work, 45 days sooner than scheduled. Then, if the Iraquis, or anyone else, releases information about this site, we will have reduced our chance of public exposure. We will have dismantled and removed all equipment in the hermetically-sealed production areas. Second, we request our government contact, to direct a reliable agency to locate Martin, gather any available data on his actions, and keep us informed. Third, a day prior to any shipment, we establish motorized patrols in security zone Whiskey, to ensure no one else sees our transport aircraft."

Major Clement's assessment triggered an extended debate, regarding several courses of action. The men argued the pros and cons. Finally, the Detachment Commander reached his decision and issued preparation orders. Then he said, "Men, let's pray our government counterpart can slow down Martin and CMR. Or else, we might all go to jail."

13

Oakland

Nolen slalomed his pickup down the winding ridge above Oakland. The bay basin lay spread before him, blanketed with the customary morning fog, hiding the low buildings, as well as the rush hour traffic. San Francisco skyscrapers poked through the vast fog bank, like lonely sentinels in a cream-colored desert. Farther west, the parabolic arch of the Golden Gate appeared to form a portal, connecting mist to mist with the sentinels on one side, and the unknown on the other.

The scene disappeared as Nolen's truck slipped into the white fog. It was time for him to begin birthing the gold rush. He stopped at a hardware store, where he purchased the necessary supplies. Two coveralls, a pullover ski mask, two trunks, padlocks, and a whisk broom completed his list. Then, he wheeled his truck to the side of the store. From a nearby dumpster, he complacently scooped garbage and paper into the cab of his vehicle. He ignored the curious stares of passing pedestrians.

His next stop was a self-storage firm, where he rented a space the size of a garage. When he located the numbered unit, he backed his gray pickup close to the side-by-side entrances. One was a standard door, mounted in the cinderblock wall. The other was a linked metal barrier, which he raised to expose the roomy compartment.

Nolen waited until he was certain no one else was in the alleyway, then yanked both trunks just inside the opening. He hastily transferred the yellow mineral into them. Using the whisk broom, he swept all evidence of sand and gold grains from the floor of the pickup. Before padlocking the footlockers, he set aside a 20-pound piece of gold, then dragged the containers to the rear corner of the storage room.

Returning outside, he removed the motorcycle mounted on the front of his truck. In one of the bike's saddlebags, he packed the large chunk of gold, Digger's notebook, and his field maps. Nolen then backed the truck into the storage room. In his zeal to be cautious, he triple-checked that nothing left inside the pickup bed could possibly be used to trace the location of his mine.

Next, he scattered pieces of mining gear around the room. Finally, while trying to hold his breath from the putrid odor of the rubbish, he scooped the garbage from the cab and layered it on top and around the footlockers. Hopefully, Nolen thought, if anyone breaks into the rented space, they'll be attracted by the valuable mining equipment, and overlook the boxes hidden beneath the offensive trash.

After stepping outside, he enjoyed a needed breath of fresh air. A satisfied smile spread across his face as he locked both storage doors. The first step of my plan is complete. Hilda and Digger's fortune is safe now. This hiding place is rented under an assumed name, a hundred miles from my home. I've paid cash for three months storage, and I'm not using the pickup anymore. It's impossible for the Army, CMR, or the police to track me down or find the gold. Nobody knows what my intentions are. And I have plenty of gold to pay for what remains to be done.

Nolen rode his motorcycle south, skirting the Oakland business district. Near the east bay airport, he selected a inconspicuous motel, where he rented a suite, using an alias.

Once inside, Nolen tossed the saddlebag under the bed and grabbed the telephone book. Noting several phone numbers, he made his first call. An assistant editor with San Francisco's most successful television station responded on the line. "My secretary informs me that you claim to have an exclusive story of national interest. We get these types of crank calls all the time, mostly from weirdos and fruitcakes. Now, I suppose you're going to give me some specific facts—something believable?" the assistant questioned, with considerable sarcasm.

"Yesterday afternoon," Nolen replied, "the police discovered a dead man in the rest room of a gas station near Susanville. I can identify that man, as well as prove why he was killed."

"You still haven't given me anything substantial. I read the overnight AP reports on the northern California homicide. Now, if our TV station was the first to identify who murdered the man..."

Nolen interrupted. "He wasn't murdered. I killed him in self-defense."

"Would you confess on camera?"

"Yes, to that, and to much more."

"Well, your information could be interesting and possibly add 30 seconds of air time. But I haven't heard anything yet that would change this rural killing into a national spread. Furthermore, by law, you're required to inform the police if you have any knowledge relative to the commission of a crime."

"I can prove the death is connected to an event that will occur within sixty days, which will create more havoc than the 1906 San Francisco earthquake. It will be bigger than the assassination of President Kennedy. Think of it—you'll know when, where, and how it will happen. You, and your station, will receive the credit for the scoop!"

"Still, just claims. I must have proof"

"Call the Susanville police," Nolen persisted. "They can confirm two facts. A metal pipe, smeared with canine blood, was found in the rest room. And the dead man worked for a large mining company. These facts would only be available to the investigators involved, or to someone connected with the killing."

"Did somebody steal something from the mining company?" the assistant editor quizzed. "Or, has something valuable been found near Susanville?

"That's all the information I will divulge at this time. Four hours from now, have your representative available to meet me in downtown Oakland. I'll inform you later where to send him. Be certain that he has the authority to bid against other TV and radio stations, for the rights to my story. Bidding starts at $35,000. Whoever wins, will win big, because they will immediately receive positive proof of the greatest news story ever sold in the history of California."

"Mister, that's much to ask for," the assistant editor replied.

"Hardly, since you'll have your proof before you pay. These days, with so much media garbage, when a truly good story breaks, the station featuring the lead captures the bulk of the viewers. If you want this story, you'll send someone. If not, you'll lose it. Remember, four hours." Nolen set the receiver down and crossed the first telephone number from his list. He repeated a similar dialogue with three other television stations, then with four radio stations.

The media calls concluded another part of his plan. But Nolen needed to plant several ideas with Hilda.

"It's good to hear your voice, Nolen," Hilda said. "I've been so lonely since you and Pal left. Are you both staying healthy? Where are you now?" She flooded him with questions.

"I'm East of Susanville, and everything's fine. Has Digger improved any?"

"I thought I felt him squeeze my hand the other day."

"Are you all right? Are you sleeping well?"

"I'm still bothered with pain, where the tumor was removed. It comes, and it goes," she revealed, with some reservation.

"Hilda, I promise when I return, I'll take you to the finest doctor in the States!"

"Nolen, do you mean you found something?"

"I've been looking real hard, but only surfaced a few ounces of river gold. I'll be in the field, steadily, for the next few weeks. I'll keep trying every day. Don't worry, if I'm out of touch for awhile."

"Son, your voice sounds a bit strange. Have you caught a cold while running around those hills? You're just as bad as Digger."

"Now Hilda, don't worry so. Nothing has happened to me that I can't handle."

Hilda's voice quivered when she spoke. "Thank you, Nolen. As always, you're so helpful and dependable. I love you."

"Take care, Hilda. If anything turns up, I'll call."

Nolen did not dwell upon the disturbing feelings caused by the first lies he had ever told his foster mother. But the falsehoods were not as important as his need to shelter her from any anxiety. Furthermore, she must believe that he was merely away prospecting, should the police or anyone else contact her.

During the coming weeks, he would need to lie again and again to safeguard himself, as well as others. He decided that he had best get accustomed to the practice.

He next dialed his older brother. "Paul, this is Nolen."

"Hey, it's great to hear from you, Nolen. When are you coming to Portland to visit? Now, that you're single again, the wife and I can introduce you to some really wonderful gals."

"Paul, I need your help—not a date." Nolen's tone was serious. "Listen carefully to what I tell you, and please, don't take any notes."

"What is it? Are you in some kind of trouble?"

"I can't provide any details now. But I urgently need you to come to Oakland tomorrow. Bring a $10,000 cashier's check. Don't tell your wife, your partners, or your secretary where you're going. Concoct some story about an emergency claim, filed against your insurance company."

"Oh sure, as if that won't arouse my wife's suspicions."

"When you get here, I'll produce proof of a lucrative deal I want you to participate in."

"Wait a minute," Paul complained. "I can't put my life on hold, without receiving a far better explanation!"

"Paul, I've never asked anything of you. This meeting is critical—both for you and for me. Trust me, I'll make you rich."

"You know, at times, Nolen, you are strange. Yes, I'll fly down. But this had better be good. I don't want to upset my wife."

"Don't worry. Just get your squatty, round butt down here." Nolen ended the conversation with a typical jousting endearment the brothers had parried between them, during the years when they grew up in different foster homes.

Nolen's neighbor answered his next call. The man agreed to maintain a close watch over Nolen's house during the coming weeks.

Finally, Nolen contacted his bank. He spoke with the manager, who had known him since he first came to live with the Jorgensons. Nolen arranged a $4,000 withdrawal from a branch bank in Oakland under the pretext of paying hospital expenses for Hilda, as well as paying bills Digger owed to bay area firms.

Nolen checked his wristwatch, then wiped his sweaty palms on his jeans. "Damn, I have to get moving." He removed the chunk of gold from his saddlebag. After wrapping it in a T-shirt, he hid it under some clothing in the back corner of the closet. As he left, he switched on the television set, and hung the "Do Not Disturb" sign from the doorknob.

Returning to Oakland, he motored to the brokerage firm where Maida Collins was employed. Inside the building, Nolen spotted the luster of her ebony hair among the clutter of people working in the large office. He silently admired her, as she conversed with a fellow broker. She sparkles, he thought. Some day I'll run my hands through her hair! Nolen paused, and shook his head, trying to control his desires. Stop that, you idiot, he reprimanded himself. Getting involved now is out of the question. Focus all your energy on your promise!

Maida's smile broadened, when she noticed Nolen approaching. "Hi," she cooed. "I was beginning to think you had forgotten me. It must be three months since we retrieved gold from my pool. How is your family?"

Nolen blurted, "A month ago, my ex-wife left to live with her parents in Virginia. A few weeks later, Digger had an auto accident, and he is presently in a coma."

"Oh, how tragic." Maida gently touched Nolen's wrist. "I'm so sorry for you and Hilda. Where is Digger? I'll send flowers."

"Thanks. It would please Digger to know you care. He's in the Susanville hospital." Nolen sidestepped from the touch of her warm fingertips. "Look, I don't mean to be abrupt, but I'm on a tight schedule. Can you handle some special transactions for me?"

"Of course. How can I help you and your family?"

"I can't discuss it here. Please, meet me tomorrow morning at 9:00. I'm temporarily staying at this address." He handed her the motel address scribbled on a slip of paper.

Maida slowly smiled. "Nolen, you certainly have a different approach for tempting a girl into your motel room." She comfortably focused her chocolate-colored eyes on Nolen.

Nolen looked away, attempting to ignore her confident, inviting gaze. "It's strictly business. After we talk, you'll agree, it's an extremely lucrative arrangement for you."

Maida did not hesitate. "I'd love to. Whenever you are nearby, my life always becomes more exciting."

14

Barracuda

17 August

In a shabby section of Oakland, Nolen considered several hotels until he found one which met his needs. Years before, it had been a fine establishment in the growing port city. Over time, the financial and business districts settled several miles away, nearer to the bay. They left behind the ten-story building surrounded by light industry, indigent neighborhoods, and a noisy, elevated highway.

Walking toward the shoddy entrance, Nolen passed several winos lingering on the sidewalk. In the dimly lit lobby, three elderly women, clutching canes and purses, ranted about the high price of macaroni. At the front desk, he rented two rooms. One was on the sixth floor, and another directly under it.

Using a hallway phone, he made follow-up calls to the radio and television stations. Only three media agencies were interested, with representatives waiting to learn more about his offer. He established the time and location for the meeting, then continued his other preparations. He rehearsed until he was comfortable with the layout of the hotel, as well as the timing of his movements. Shortly after he concluded his arrangements, a knock sounded at his door. Nolen slid on a ski mask, then opened the door.

An attractive, short woman with red hair abruptly stepped back, shoving her hand into her tailored coat pocket.

"It's safe," he said. "I'm wearing this hood merely to conceal my identity. Only the person who wins the bidding sees my face."

"Has anyone else arrived?" she asked.

"Not yet. They should be here any moment. Come in."

"How often do defenseless women walk into your hotel room, when you're wearing a mask?" she challenged. "I'll wait here, until someone else shows."

As she spoke, the elevator door opened. A San Francisco TV reporter, whom she recognized, strode down the hallway. "I take it this is the mystery caller," the man remarked. "Are we meeting out here?"

The woman released the item she clutched in her pocket, and pushed her way past Nolen. "The name's Pam Pendleton. I represent TV station 25 in Oakland."

The two early arrivals attempted to pry information from Nolen. He ignored their inquiries, telling them to wait until the last person joined the group. Finally, a newsman from a San Jose radio station entered the hotel room. Nolen verified each identification, to ensure they were in fact news representatives from the appropriate agencies. Then, he summarized the ground rules for all three, emphasizing that the highest bidder would receive the story, together with positive proof, prior to signing any agreement for monetary payments.

"Your hints of a valuable story provide no basis for my station to offer you thousands of dollars," the San Francisco TV reporter bluntly argued.

Nolen hesitated, then said, "The autopsy will have verified that traces of gold were on the hands of the dead man." Silence filled the room.

"Thirty-five thousand dollars." Pam interrupted the silence with her husky voice. The men stared at her. She lit a cigarette. "You're right. It is too risky a story. Good reporters have been fired for less. But I have an iron-clad contract with my station. So, why don't you two allow me to take all the chances?"

The radio representative scowled at her, and promptly raised the bid by $500.

"Thirty-six thousand," the male TV reporter angrily joined in.

"Forty," Pam responded. "Both of you should just forget this story. I doubt that either of you could handle it anyway."

"Forty-two thousand," the radio station representative countered.

Pam casually opened her compact mirror and wiped a speck from her penciled eyebrow with her pinkie finger. "Forty-five thousand," she commented, as she snapped her compact shut.

"Screw you. Fifty thousand," replied the other TV reporter. The price climbed higher, in leaps of one or two thousand, as the bidding escalated between the two men, until it peaked at $63,000. The four glanced back and forth at each other.

The woman spoke again. "An even seventy thousand."

"Damn you, Pam," the other TV reporter growled, raising his voice. "You can have it. I'm not about to offer that much money without better assurances. I doubt his story will be as valuable as you apparently think. You're grasping at straws. It won't help you out of your situation."

"What do you mean?" she snapped.

"It's well known, among the local media, why you were removed from prime time. You were pushed aside for a more personable, younger woman. One who could increase the news ratings. Do you really believe you'll get back in front of the lights through this gamble?"

Her eyes narrowed as she responded, speaking with an exaggerated southern drawl. "Why, hoooneeey, I dooo believe you are upset with little ol' me, for winnin' this in-sig-nif-i-cant story, with my most superior chaaarms. I sooo enjoyed the look on your face—when I beat your paltry bid!"

The San Francisco newsman turned to Nolen. "Buddy, in any deal there's two things to consider: how much profit you make, and the integrity of your business partners. You can call around town and verify that if you sell to me, you will get paid. You'll never have to worry about being stabbed in the back. Those same people will also explain that you're about to make a big mistake. Pendleton here is a female barracuda, known to have a vicious bite. You'd get less money teaming with me, but you would be a whole lot safer."

"I don't have the time to fool around," Nolen answered. "I'll take my chances."

"Well, buddy, I'm not letting this mystery elude me. Checkbook journalism isn't the only way to obtain a story."

"Take your sour grapes and leave," Pam retorted. "I have business to conduct." She tapped the ash build-up from her cigarette with her polished, red fingernail and waited for her competition to leave the room.

Nolen closed and locked the door, then removed his hood. He tossed it onto the bed as he strode to the window. "I thought you reached your bidding limit when you went quiet, after offering $45,000."

Pam paused, and slowly crushed her cigarette in an ashtray. "It happened long before that." She looked into his startled eyes. "I'm only authorized to pay $38,000."

"What? You mean you can't pay the $70,000?"

"Certainly, I can arrange it. But only if the story is worth it. Otherwise, we renegotiate."

"No way," Nolen angrily retorted. "I can still get one of the other bidders to buy the story."

"Don't bet on it. If I walk out of here complaining, they'll think something's wrong, and will drastically lower their bids. Anyway, it's apparent you're trying to captivate a national news agency with your story. Well, you've got one. Now, your information must stand on its own merit, before we agree to a price. Screwballs are always striving to peddle some scoop. To protect ourselves, we're careful to pay only for newsworthy information. Where's your proof?"

Nolen stared at the woman, assessing the unexpected development. "Lady, you might appear small and frail, but you're no lightweight. You are a surprising combination of brains and guile. Just the type of person I'm going to need, even if that TV rep was right about you. Come with me." He raised the window and stepped onto the fire escape.

She hesitated, then shrugged, as Nolen extended his hand to assist her down the rusty, iron stairway. After crawling through the window into his second room, Nolen pointed to gloves and coveralls laid across the bed. "Put these on. Be quick about it."

"Wait a minute. Tell me what we're doing. You're involved in a murder, and acting very strange. How do I know you're not some psycho?"

"Look, my intent is to leave this hotel, without those other reporters tracking us, hoping to scoop this story. Your competitor said he wasn't through yet. For all we know, he may have even alerted the police. So, these clothes will help disguise us."

She held the coveralls against her petite, 120-pound frame. "Anything smaller in your wardrobe? You can bet I won't win any fashion awards wearing this chic ensemble."

Nolen finished zipping his suit. "No. That little number isn't furnished in varied sizes. I wasn't expecting a petite woman would win the bid. Roll up the pant legs and sleeves, and let's go!" He opened the door, peering in both directions, checking the hallway.

Pam took the opportunity to transfer her .25 caliber pistol from her blazer jacket into the pocket of her jumpsuit.

Assured that the corridor was empty, Nolen led the newswoman to the freight elevator, where he punched the call button. He glanced up and down the hallway, and again stabbed the button several times. Finally, the door to the service elevator creaked open, releasing a strong odor of stale cigars and disturbed dirt. They descended to the basement garage, where his motorcycle stood ready, next to the elevator.

"Seems you've carefully considered every move," she observed.

Nolen handed her a helmet. "Maybe, but we're not in the clear yet. When we leave, see if those other two guys are out front, or if anyone follows us."

The motorcycle roared to life. He wheeled it through the dark, damp parking area and up the ramp into the bright afternoon sunlight. They slowly rode past the hotel lobby, allowing Pam to survey the street. The emergency lights of a police car flashed, as it sat parked in an alley next to the building.

"I don't see anyone I know," she yelled in his ear. "You were right, though. One of my competitors must have told the cops about our meeting."

Nolen chose a meandering course, doubling back on his route several times. When satisfied that they were not being followed, he stopped the bike in a store parking lot and got off the motorcycle. "Put this blindfold on."

"What the hell for?" the newswoman snapped.

"I don't want you to know where my proof is located. It will only take about ten more minutes. Please, cooperate."

"No." Her voice sounded firm. "Not until you give me some reason to believe you have something."

"Look, during the last six months, my foster father and I have been searching near Susanville for gold. Three days

ago, I discovered a large vein. Yesterday, while I was returning home, two men employed by Continental Mining and
Refining attacked me. They killed my dog, and attempted to
steal the ore I had extracted. Here are their wallets, with
identification cards, and several receipts establishing their
connection to CMR."

Pam studied the evidence. "Well, your story tracks with
what the Susanville police said. In fact, the officer I questioned was surprised that I knew about the dog's blood on
the pipe, as well as the involvement of a mining company.
They confirmed the homicide victim periodically worked for
CMR. After my chat with the police, I did some research on
CMR. It's the third largest mining company in the U.S. But so
far, you've described a mildly interesting attempted robbery
which appears to have gone wrong. Where's that national
importance you advertised?"

"I've discovered the richest gold strike in the history of
California. It's bigger than any discovery made in 1849,
when people spent months traveling from the east coast to
reach California. Imagine the magnitude of this gold rush,
when you present the world television coverage of my as-
tounding discovery. This time, the gold rush won't take
months to materialize. Overnight, thousands of people from
California and the surrounding states will flood into
Susanville."

Pam considered his comments. "It still sounds like a preposterous con."

"I can prove I'm telling the truth, if you'll trust me for just
ten more minutes. Besides, you can keep your hand on that
pistol you have tucked in your pocket, so you will feel safer."

She paused, surprised at his knowledge of her weapon.
"OK, I'll do it. And you can bet I'll use my gun, the very moment I feel you are trying something suspicious."

She tied the blindfold across her eyes. Nolen verified it
was secure, then helped her put on the bike helmet. He

flipped down its shade visor for additional assurance that she would not memorize the route to his gold. Again, he followed a serpentine course. But this time, he endured the hard nose of her pistol jammed against his spine.

After a quarter of an hour, he parked in front of the rented storage area containing the trunks. While Pam stood by the bike, Nolen opened the small walk-in door, and pulled one footlocker toward the wedge of sunlight spilling through the doorway. Then, he led her inside the storage room. "Now, remove your helmet and blindfold."

She followed his instructions. Nolen watched her reaction as the coverings were removed. "Damn, this place stinks! Where are we?" She covered her nose in an attempt to avoid the foul odor, as she looked around, blinking her eyes, adjusting to the light.

She saw the pickup, camping gear, rubbish, and prospecting tools, then glanced down at her feet. She gasped when she realized she was looking into a trunk filled with gold. Kneeling, she dug through the box, verifying that no false bottom existed. "I want an assay," she whispered.

"You choose what you want, and I'll chip off a piece for you."

She selected a rock the size of an egg and turned around. "This will do." Pam squinted, as the bright light shocked her eyes. She shaded her face from the reflection bouncing off the blue and orange striping, covering the wall of the storage bay across the driveway.

"OK, take the whole thing. Get it analyzed. But don't tell the assayer, or anyone else, where it came from! Then, as security on our deal, put the gold in a safe deposit box. By noon three days from now, I want two cashier checks, each for $35,000. One, I receive that day. The second, you hold until I submit the claim in a month. Until then, you work with me, recording what happens, and preparing your news releases. How does that sound?"

"Weird, that's how it sounds. I'm wondering, why are you selling me this story when you have a fortune at your feet?"

"I'll explain everything tomorrow morning." Nolen assured her.

"Well, I must inform my boss. This story has wondrous potential."

"Pam, it isn't in your financial interest to do that. Do not tell him anything about the gold, until after you get the sample assayed, and talk with me tomorrow."

"Are you going to cut me in on this, now that I know about your strike? Then, I could purchase Channel 25 and fire the bastard who demoted me!"

"Tomorrow I'll explain the offers I'm making."

She continued to fondle the gold rock as she spoke. "Sure. Take me back to my car, so I can get started right away."

"No. I'll take you to a taxi. Stay away from your car, house, and office for a few days. It's possible you'll be followed by another news agency."

"Damn, you're right. H'm...I suppose I can work within those constraints. Where can I contact you?"

When Nolen was satisfied with all of the arrangements, he again blindfolded her and led her from the storage area. Within minutes, Nolen drove several miles from the rental space and secured a cab for Pam.

15

Partners

Tower appeared posed for a portrait, with his dove-gray hair and professionally tailored suit, as he sat on the edge of his office desk, fingering his vest chain. He impatiently watched his cousin, Boodan Tribou, thumb through the confidential file.

The Cajun was a compact man, powerful and sinewy. He was toughened by thousands of hours of outdoor living and hunting other men. His skin favored cracked leather, reddish-brown in color. Prominent veins snaked over knotty muscles in his forearms. A decade earlier, a knife fight on an oil platform off the Louisiana coast had cost him his left eye. Since then, he proudly wore a custom-made, red patch over the injury, as if displaying some badge of valor. Jet black hair fell across his forehead, when he turned his face to better view the report with his one good eye.

Looking up at Tower, he questioned, "What exactly do you want from me, cousin?" The tone of his voice was deep and sullen.

"I want you to find Martin, as well as determine the location of his strike, before he files an official claim."

Boodan peered at Tower. "You mean, steal his claim."

"I did not say that. However, I am attempting to ensure that CMR files first. Then, any subsequent challenge by

133

Martin would be suspect, should we ever go to court. You have done jobs like this before. Just do it again."

"You understand," Boodan emphasized, "I hire and fire the friggin' people on my team? And I choose how to do the job. I don't want none of your mousy execs getting in the way."

"That sounds reasonable. None of the managers here could do what's necessary, anyway. However, remember, I'm in charge of this operation, and I expect to be informed before you employ any extreme measures. You have a reputation for being, let's say, excessively forceful."

"Shit, like you don't."

"Such an approach may be appropriate under certain circumstances. But here excessive directness has the potential to damage me. So, I'll decide when to use force. Agreed?"

Boodan tossed the file onto Tower's desk. "OK, as long as you remember, people get pissed when I make 'em poor."

"I understand you will not be subtle; just control it."

"This Martin character's been alerted that we're after him. He's dangerous now, so I want some big bucks for this job. CMR pays all bills, including providing a company jet to fly me around, whenever I want. Plus, I get paid fifty grand, whether or not I deliver the information you need. When the strike is filed, I get ten percent of the mine production."

The haggling began. During the ensuing half hour, Tower twice cajoled Boodan from leaving the office. Gradually, the cousins came to an agreement. CMR would cover all expenses of the team. In addition, the Cajun would receive $40,000, plus two percent of the mine profit, if he were successful.

"How soon can you begin?" Tower asked.

"I began before I left Texas. I told your assistant to get that slimehole Daniels, the one who botched the kidnappin', flown down here for me to question. He may know something valuable. I also brought six experienced men with me.

I'll take a couple to Monterey this afternoon, to Martin's home spread. We'll see if he left a trail for us to follow. If he's like most dumb prospectors, he'll high-tail-it back home, drink too much whiskey, then brag about his strike."

"I doubt that we will be that lucky with Martin."

"Two of my team will head for Susanville, to see what they can learn. I'll send the others to Reno, to set up our base. From there, I'll finish fleshin' out the crew, to ten or fifteen people. I'll give you a status report every day. That soon enough?"

Tower chuckled. "Boodan, I admire how you always cut through life's bullshit. Coordinate all the details with my assistant. You can reach me any time through this office."

The patron and the tracker stood together and shook hands, sealing their bond of destruction on another man's life.

That same afternoon in downtown Sacramento, a secretary ushered Cholo Cantera through the opened doorway of an austere office, in the one-story Federal building. As the tall young man strode beyond her into the room, the woman peripherally observed him, admiring his high cheekbones and caramel-brown skin. Only a faint scar on his lower lip blemished his otherwise stately and clean-cut features.

"Agent Cantera reporting, sir."

The Special-Agent-in-Charge reached across his desk and extended his hand. "Cantera, glad to see you. I understand you recently arrived from Quantico, Virginia?"

"Yes, sir, I just graduated from the FBI academy, two months ago." An irritating trickle of sweat wiggled down Cholo's neck, as he stood in front of his white-haired superior, noticing the hardness in the older man's face. Cholo recalled the stories his new teammates had revealed about the SAC, when they learned Cholo was to report to the front office.

Several agents claimed their boss was the toughest, most demanding bastard they had ever worked for. Simultaneously, they felt he was one of the best crime fighters in the Bureau. Before his appointment to Sacramento, he had been on special detail to the White House Drug Prevention Committee, then chaired by the Vice-President. All the agents speculated as to why the SAC was farmed out to a regional office, after such a favorable assignment. Several agents postulated that his ample skills must have been recognized by a powerful mentor, but he may have stepped on some political toes. Otherwise, he would have been assigned to the Washington D.C. headquarters. In any case, everyone warned Cholo to keep his mouth shut, do whatever the SAC ordered, and work diligently.

The senior agent perused the open file on his desk. "Your team leader recommended using you, Cantera. You were third in your Academy class. That's commendable for a minority."

"Yes, sir." Cantera masked his feelings when he heard the minor slur. He had learned to control his emotions, as a youngster growing up in southeast Los Angeles. Subtle and blatant prejudice against Mexican-Americans was something he had continually experienced over the years.

"Son, you're qualified, and you're not presently assigned to anything important. So, I'm transferring you to work for me on a special case. Any objections?"

Cantera's spirits rose. The offer was something that rookie agents dream about. Carefully, he gave his reply. "Sir, your reputation is well-known throughout the Bureau. I'm honored to work for you."

The complimentary statement caused the senior agent to smile. He pushed a thick folder toward the young man. "You'll be my 'go-fer' in the field. I expect you to locate this guy, Martin. Inside the file, you'll find a special letter of authorization. It will ensure you receive cooperation from any

federal or state agency that you contact. If you have any difficulties, call me."

"Yes, sir."

"Realize that this case is TOP SECRET," the SAC dictated.
"You report only to me. Tell no one else what you are doing.
I mean, no one! Furthermore, I expect to be briefed regularly
regarding all that you uncover."

"I fully understand, sir." Cantera tugged at his tie which
seemed too tight.

"My secretary will provide you a work room, nearby.
Study the report, then see me afterward. I'll lay out your investigation plan at that time. I want you in the suspect's
hometown by tomorrow morning. Any questions?"

"No sir."

"Get to it."

A few minutes later, seated at a small table in an empty
office, Cantera opened the dossier. He began reading the
case fact sheet.

1. <u>Mission</u>: Locate, follow, and report the activities of
Nolen Martin, SSN 532-41-0038, regarding federal felony incident 1037-99-CID36.

2. <u>Intel</u>: Subject is a member of an ultra-right-wing
group, involved in theft of Army weapons and related
ammunition for use against the U.S. government. Martin has extensive knowledge of terrorist tactics. See red
tab, military records, Annex 3.

3. <u>Felony</u>: At or about, 0315 June 16 of this year, one M16
rifle and twenty-four 4.2-inch, high-explosive mortar
rounds were stolen from the U.S. Army at range 10A,
Camp Hunter Liggett, California. One military guard
was assaulted by two men. (See enclosure 1, Criminal
Investigation Division Case 1037-99-CID36). Weapon
and rounds were taken from the overnight storage site

at the range. This activity violates federal laws pertaining to theft of U.S. government property.

4. Suspect:

    a. Location: Residence, 210 Nueva Drive, Monterey CA.

    b. Description: Male Caucasian, age 35. Height 5' 8", weight 180 lbs. Build medium, hair blond, eyes green. Photograph in Annex 2.

    c. Physical Condition: Unknown. Military medical records in Annex 4 indicate at time of discharge, Martin was in excellent health.

    d. Mental Condition: Unknown, assumed stable.

    e. Political Orientation: Since being released from the U.S. Army, the suspect has raised funds for the ultra-right wing political group: White Supremacists.

    f. Personal History Annexes:

#1: Birth Certificate	#2: Photograph (U.S. Army)
#3: Military records (U.S. Army)	#4: Medical records (U.S. Army)
#5: Education records	#6: California driver's license
#7: IRS records	#8: Credit reports

The young agent sat back in his chair, noticing his excitement. He realized the need to calm himself, to think clearly. He did not intend to make any mistake. Cantera rarely blundered because he had received eight years of invaluable training prior to attending the FBI academy. He had received that special tutoring during high school and college, from his uncle.

During the last year of J. Edgar Hoover's reign, his uncle had joined the Bureau. The old Mexican-American was one of the first minority agents in the FBI. As such, he had not been appreciated by many of the senior agents in the Bureau. Often, he was forced to deal with elitist agents, who did not accept hiring females and non-whites. Nonetheless, he produced outstanding results, gradually rising within the hierarchy, and ultimately becoming a legend.

Cantera's uncle was proud of his contributions while in the FBI. Repeatedly, he bragged about the Bureau's heritage, traditions, and the many good men and women who sacrificed for their country.

Over the years, he shared many case experiences with his nephew. As Cantera's interest in law enforcement grew, his uncle spent hours teaching him crime detection and street-survival skills. The uncle was a dedicated instructor. He wanted his sister's boy to become a successful agent, and avoid suffering through the same mistakes that he had made during his career.

To train the boy, Cantera's uncle would describe a crime, explaining the available clues. Then, he would challenge young Cantera to determine what action should be taken, where to look, whom to talk with, and what questions to pose. When the boy was correct, he praised his efforts. When the boy erred, he would detail better ways to solve the case.

By the time Cantera reached the academy, he held a distinct advantage over most of his classmates. However, unknown to Cantera, his uncle had his academy friends promise to be particularly critical when evaluating his nephew. That process further honed Cantera's skills, but it also kept him from becoming the top graduate in his class.

After Cantera relaxed, he listed the facts that he knew, as well as the assumptions, and his questions. Countless times he had been drilled on what to look for in a well-prepared report. Why, he wondered, doesn't the summary describe what investigations had been accomplished on Martin, during the months since the incident in Monterey county? And nothing in the file explains how the suspect was connected to the theft. Such information should not be missing from a case report.

Is the SAC merely testing me? he pondered. Or perhaps, real field reports do not always attain Academy standards. He began speed reading the annexes. Other doubts troubled him, as he worked his way through the file. This suspect seems too straight to be involved in stealing government property. And why are IRS records included in the file? Those are only released when federal banking laws have been violated.

Again, Cantera ignored his doubts. After outlining an investigation plan, he returned to his supervisor. The aged agent agreed to several suggestions, while pointedly dismissing questions regarding any file discrepancies.

16

Association

18 August

"Damn wind," Pam Pendleton swore, as she strutted into Nolen's motel room. "These morning gusts off the bay always muss my hairdo." Preoccupied with her appearance, she settled at the kitchen table, and checked a compact mirror to smooth her red hair. Satisfied, she flipped open her notebook and surveyed the suite, jotting her first impressions of the meeting.

It was a small room, with walls too thin to muffle the traffic noise, reverberating from the nearby interstate highway. Numerous nicks and smudges, from chairs and luggage shoved carelessly about by previous occupants, marred the faded, cream-colored walls. Plastic vinyl improved the appearance of counters and tabletops. But upturned corners and gouges hinted at tackiness and a lack of timely maintenance.

It's a favorable sign for me, Pam thought. My best stories always seem to originate in some dive, like yesterday's hotel and now this dump. She focused her interest on the latest arrival, a lithe, younger woman with raven hair. Pam caught the look of restrained warmth between the newcomer and Nolen, as he helped the woman remove her light wrap.

Nolen appreciates the gawky girl a bit too much, Pam postulated. Are they more than just friends? Do they sleep together? And why not? Wouldn't anyone who beds Nolen,

also bed his fortune? Well, I won't let some fresh-faced innocent steal this opportunity from me! She's probably just another young slut, like that bimbo at the TV station who
robbed me of my anchor seat. I'm not about to lose out again!

Pam took a sidelong glance at the potbellied man seated
across from her. Why, she wondered, do balding men persist
in emphasizing their lost youth, by draping two or three
long stands of hair across their shiny domes? How ridiculous. He looks like the proverbial traveling salesman, with
jiggly jowls and perpetual beads of sweat accumulating on
his upper lip. There he sits, too warm, wearing a suit which is
too tight, proving he is out of shape and overweight. But he
won't slip his jacket off to be comfortable, because he believes that would be inappropriate. It amazes me, how oblivious men seem to be regarding their appearance. Some
awareness, coupled with a little effort, would geatly im-
prove their attractiveness. I'm glad I attended finishing
school to learn how to maintain a polished look.

The salesman drummed his fingers upon the table, in an
irritating fashion. He waited for the new arrival to be seated,
and for his brother to begin explaining what was so important. "Nolen," the heavy-set man said, "I hope you haven't
drawn me into some stupid venture. You always choose to
ignore the risks, always pursuing something radical."

"Have I ever led you astray?"

"More than once. And when I return home, I'd best have
a believable story to settle my wife. She wasn't understanding when I explained how important it was for me to fly
south to see a client, who just suffered a major fire loss.
That's what twenty years of marriage will do. It creates
built-in radar in wives. They instinctively recognize when
their husbands are doing something they shouldn't."

The group of strangers scanned one another, feeling the
awkwardness that permeated the room, aware of the unusual purpose for the gathering.

"Now that everyone has arrived," Nolen said, "I'll introduce you." He placed his hand on Maida's shoulder, gently squeezing her in recognition. "This is Maida Collins. Maida is a super stockbroker from Oakland."

Maida's many black curls bounced about her neck and shoulders, as she flashed a smile at Nolen. "I'm glad you feel I'm so talented."

Nolen pointed to the balding man. "And this is my older brother, Paul. He sells insurance in Portland, Oregon, and has a genuine knack for putting deals together. Paul hasn't yet achieved his life's goal of becoming a millionaire."

"I don't want much," Paul interjected. "Just enough to cover life's trivial necessities, like an income of $100,000 a month."

Nolen continued. "This attractive lady, across from Paul, is Pam Pendleton, a television reporter employed with Channel 25 in Oakland. Pam has a natural intuition for breaking exceptional news stories. She is continually striving to scoop her bay area competitors."

Pam responded. "I wish you would get on with the agenda. It's time for you to provide us with some facts on what this meeting is about. After all, you tweaked my interest yesterday, when we last met." She barely smiled at Maida, intending that the younger woman should recognize that she had some significant competition.

Nolen sensed the group's watchfulness, as he removed from the closet the large rock covered with his T-shirt. "I realize you all are anxious to learn why I asked you to meet with me today. Now, I'll show you." Nolen smacked the stone onto the table surface. "I discovered a fabulous gold vein!" He jerked the T-shirt from around the rock.

The group reacted. Eyes expanded. Jaw muscles tightened. Maida uttered a distinct gasp. She stared at the ruby-yellow ore, appreciating its value, exceptional by comparison to the much smaller nuggets extracted from the pool near her cabin. Maida flipped a strand of hair off her shoulder, then reached forward to touch the gold. She pulled the rock toward her for closer examination, amazed by its weight.

Nolen spread a document on the table. "This assay, performed in Susanville, proves the extreme purity of the gold."

"How much is this worth?" Maida questioned, unable to quit staring at the precious chunk of gold. "Ah...how big is the vein?"

"In my estimate, the mine can produce three tons of gold during the first week of operation. That's worth thirty-six million dollars at today's prices. It is the richest mineral strike in the history of California."

"Incredible—positively incredible!" Pam muttered, shaking her head in amazement. She lit a cigarette in an attempt to conceal her excitement.

"But why do you want us?" Maida asked.

"I need your help to successfully bring in this mine. Prospectors have found gold throughout history. In most instances, other people, in one way or another, have taken their wealth. Claims are jumped. Miners are murdered. Crooked judges steal the rights to the gold. Partners swindle each other. Or, governments expropriate the mines. To by-pass those misfortunes I have devised a plan to protect this mine, and all of us as well."

"If this vein is as valuable as you claim," Pam responded, "it had better be an exceptional plan."

Nolen continued with the details, mesmerizing them with his vision. "Each of you will earn five million dollars for your efforts during the next sixty days. Plus, millions more, when the mine becomes fully operational. But you only receive your share if you work together, and do what I tell you, when I tell you."

Paul scanned the assay report on the table. "Nolen, are you certain you're not exaggerating? How do we know this assay is correct?"

Pam snatched the sheet from Paul's hand, to study the report. "It's accurate," she confirmed. "Yesterday, Nolen gave me a sample of the ore. I took it to a chemistry professor friend, at the University of California. He was most inquisitive as to how I had acquired a lump of gold worth $10,000."

"Why did she know about the mine before your own flesh and blood?" Paul challenged.

"Simple," Nolen answered. "An unbiased source needed to verify the mineral content of the ore. I felt all of you would more readily trust me, if you had a second confirming assay. OK?"

"You didn't answer my question," Maida persisted. "What is it that you want from us?"

"We're going to form an Association, to build a boom-town in front of the gold rush. Imagine the wealth people could have earned in 1849, had they known a huge gold rush was imminent. For very little risk, they could have made a fortune by selling goods and services to the prospectors who raced into the gold fields. Today, we four are the only ones in California who know that, after the gold strike is announced, thousands of men and women will rush in like locusts. Our Association will reap the money that those people carry into the gold field, many truck loads of money."

"How the hell are we going to do that?" Paul questioned.

"I will control fifty-one percent of the mine," Nolen responded. "Each Association member will be guaranteed an eight percent share of mine revenues. The remaining percentage of the mine ownership will be reserved to buy the people that we'll need to assist us."

"What must we do," Pam questioned, "to earn our eight percent?"

"Each of you will manage a primary boomtown project, where you will employ the special skills you've acquired over the years, working in your profession. You'll receive a percentage of profit from each Association project. However, you'll acquire a much higher share from the assigned project that you direct. The better you manage, the more you make."

"Why must we accomplish these tasks within sixty days?" Maida asked.

"Word of a strike could leak out at any moment. Continental Mining and Refining, the third largest exploration company in the U.S., knows that I have made a big discovery. They attempted to kidnap me two days ago, and now they are searching for the mine. We must file a claim to protect the site, before CMR stumbles onto it."

Paul frowned, as he wiped unexpected sweat from his hands. "Kidnapped? What do you mean, kidnapped?"

Nolen related the incident at the gas station.

"What if something like that happens to you again?" Paul persisted. "What if you're not so lucky the next time? If CMR is willing to kill you, they surely would jump on us after you're dead. By joining in this scheme, we also expose our families to danger and financial ruin!"

"The boomtown is your protection, should we somehow lose the mine. It gives us two shots toward making a fortune, rather than one. And, rest assured, I want to remain alive. So, whenever possible, I will keep traveling around the state, to prevent CMR or anyone else from locating me. You three, plus a few others later on, will accomplish most of the project tasks, following my guidance."

"I knew this would be risky," Paul said. "I knew it."

"Furthermore, should I die before the filing, none of you will receive anything, because I'm the only person who knows the location of the gold ledge. The best way to ensure that we file, is to never tell anyone where I am, or divulge my activities. It's in your best interest to make certain no one, within this group or any outsider, harms me. Furthermore, should I disappear after officially placing the claim, there will be no increased benefit to the Association, since my shares will go to an undisclosed, silent partner."

Paul's face turned pale. "You're crazy! The sun must have fried your brain. Or you finally fell prey to your foster father's wild dreams."

"Paul, you, more than anyone else, should appreciate these incentives and safeguards. They're my insurance for staying healthy."

"You're proposing to build an empire overnight, fight a powerful corporation, while placing all our lives in jeopardy! With what? Look around you. This isn't the Rockefeller Foundation, nor the board members of Standard Oil. We're just a bunch of ordinary people."

"He does have a point," Pam agreed. "You're asking us to accept many, many risks. We need more assurances."

"Think for a moment," Nolen urged. "Isn't this offer what every one of you has dreamed of, ever since you were a kid? Isn't this the challenge of a lifetime? A chance to make a fortune, to become famous? The opportunity lies before you. All you must do is trust in yourselves. The only difference between people like us and those controlling major corporations, is that when they saw a great opportunity, they seized it."

"It can't be done!" Paul disagreed.

"We have the initiative," Nolen replied. "Realize the significant advantage that provides. No one knows what we are attempting. By the time anyone discovers our strategy, they won't be able to stop us."

Maida voiced her concern. "We will certainly need more than initiative to be successful."

"Don't underestimate your capabilities," Nolen urged. "Each of you is shrewd, well trained, and capable. Haven't you, at some time or another, complained about how poorly your bosses were operating their respective companies? Often, you recognized significant problem areas, then conceived excellent solutions. But you didn't have the authority or position to implement any changes. At this moment, I'm offering you that power!"

He opened a cabinet and removed a stack of folders. After handing one to each person, he pinned a large chart to the wall. "Study this task network, together with the accompanying instructions. When you finish, you'll have a concise picture of the primary project assigned to each of you. You will understand what projects complement each other, when you must accomplish certain tasks, as well as your profit share of each business. You'll see how feasible it can be, to become incredibly rich, within a very short time span! Just evaluate my plan. Then, do what you have always dreamed of doing."

Nolen stepped away from the wall, encouraging the other three to scrutinize the partnership diagram. Their concentration was intense, as they began comprehending the magnitude of their roles, and a future they had never anticipated.

Gradually, the salty odor of perspiration tainted the room. Pam bit her lower lip, lit another cigarette, then abruptly stubbed it out. Paul continued to mop sweat from his forehead with his handkerchief. Maida shifted her weight from one leg to the other, repeatedly twirling a coil of black hair around one finger.

"This so-called plan of yours is merely an outline," Paul blurted.

"And a damn sketchy one at that," Pam added.

"That's correct," Nolen acknowledged. "I don't have all the answers now. You will fill in the outline with detailed planning, then implement your portion."

"It's not a sure thing," Maida stated.

"There is some risk, even though the plan is simple," Nolen confirmed.

"The stock manipulation isn't totally legal either," Maida remarked. "Are you aware of that fact?"

"We won't harm anyone by making those purchases. It's illegal only to the extent that we are not informing the SEC. Be extremely careful about who learns of our involvement with those purchases."

After totaling the potential millions he would earn in the next year, Paul turned from the wall chart. "Despite my skepticism, Nolen, I suppose I'm capable of doing all of the tasks you list, including flying to the Caribbean next week. But I've never handled some of these types of transactions. And your schedule is tight as Hell. It demands a tremendous amount of coordination."

"Just be creative," Nolen assured Paul. "I'm confident you'll get the job done."

Pam pointed to the wall chart. "Your plan requires us to work with different people to perform several different projects. Our responsibilities are split. Why?"

"It's a simple checks and balance system," Nolen explained. "I designed it to prevent someone's greed from surfacing and causing problems. We are all cross-teamed, to ensure that no partner successfully steals any lucrative project from the Association, before the rest of us learn of the attempt. Someone with interests other than yours, will continually be looking over your shoulder. You gain by pulling together, not by pulling apart."

Maida inquired, "How will we finance this plan?"

"We establish a fund," Nolen elaborated, "with each of us contributing as much as we can. For every start-up dollar you provide, the Association will pay you back six dollars. I know, approximately, what each of you can afford, and I have estimated each minimum contribution. Paul, you should be able to provide $50,000, Pam $40,000, and Maida $20,000. I'll donate $39,000."

"Wait a minute," Paul snapped. "Where did you get that much money? You told me just a few weeks ago, with your divorce expenses and paying your foster folk's debts, that you were practically broke!"

Pam snapped her fingers. "That's why you sold the news rights to my station. You needed cash to help start the Association rolling."

"Yes. Selling the TV rights to our inside story, about the gold rush, provided $35,000 of start-up capital. And, I'm including an additional $4,000, which is every dollar I had in my savings. I'm also contributing a trunk containing 245 pounds of gold. So, our initial working capital will total $149,000 cash and one and a half million dollars worth of gold ore. OK, Paul?"

"Amazing—friggin' amazing!" his brother muttered, while shaking his head in awe.

Nolen continued answering questions and addressing concerns. Slowly, the probing questions evolved into suggestions for improving his original plan. Finally, Nolen requested commitment. "Will you do it?" Pam and Maida responded positively. Only Paul hesitated before agreeing.

"Great," Nolen concluded. "Last night, I drafted the partnership contracts. Each of you must sign them before leaving today. But now that we've taken our first steps toward achieving our fortune, let's celebrate."

Nolen retrieved two bottles of champagne from the refrigerator. The men and women toasted the formation of their new partnership. Their four glasses clinked in unison, while each wondered how drastically their lives would change, during the months to come.

As they continued discussing the many intricacies of Nolen's plan, they crosschecked, sharing suggestions regarding how to handle the special projects. They considered what problems they might meet and overcome. Increasingly, their excitement and anticipation over great wealth suppressed the worrisome possibilities of failure. Nervous laughter helped squelch genuine concerns. Their initial, tentative steps intertwined their hopes and destinies, and moved them toward an uncertain future.

Maida scribbled her home address across the reverse side of her business card. She slipped it into Nolen's hand and quietly whispered in his ear. "I need to talk to you—alone. Can you see me tonight?"

"You certainly have a different approach for inviting a guy to your home."

Maida smiled. "No, it's just business, something that I'd rather not discuss here."

"OK, I'll try to arrive before dark."

An hour later, the meeting ended, and Nolen was alone in his suite. In front of the group, he had acted calm, confident, and controlled. Now, a burning knot throbbed within his belly, as he lay on the motel bed. Troubling thoughts, concerning the day's events, disturbed his concentration.

I wonder if I can actually hold it together long enough to survive? Today's outcome went the way I planned, even though Paul was shakier than I anticipated. I'll have to keep him away from dangerous situations. The others will be scattered all over the West Coast. Someone might make a mistake or break down under pressure. Is Paul right? Maybe, this effort is too difficult?

Nolen gnawed at one of his fingernails while his thoughts rambled. When he recognized his tension, he deliberately took charge of his emotions. His determination and defiance returned. "Screw it! Somehow I'll make it work!"

However, the unrelenting pain in his stomach continued.

17

Night Moves

18 August

At dusk, Nolen rode his motorcycle past Maida's hillside condominium, carefully observing the activity in the neighborhood. Young professionals populated the up-scale district, maintaining landscaped lawns, flower beds, and minivans with personalized license plates.

Nolen needed to verify that only Maida was waiting for his arrival. Passing by again, he ensured that her open garage, as well as each vehicle parked along the street, was unoccupied. He stopped uphill from the residence. For twenty minutes, he watched for anything suspicious. Only a man and woman moved about in their yard, playing ball with their youngster. When Nolen was satisfied that it was safe, he rolled his motorcycle down the hill, parking next to Maida's polished BMW.

She opened the door to her kitchen, as he removed his helmet. "Would you care for a drink? I suspect you need one."

"A beer would be great, lady."

She selected a chilled bottle from the refrigerator and handed it to him. "Please, come into the living room. No one else is in the house, so you can relax."

Maida's subtle perfume tantalized Nolen, as he followed her into the main room. While she opened the window blinds, to gain a better view of the sunset, he settled onto a couch. He appreciated her narrow hips and wavy hair, interestingly coiled above one ear. As she walked back toward him, he noted how the skillful application of her makeup highlighted her cheekbones and enhanced her expressive eyes.

"Isn't the view of the valley exceptional?" she inquired.

Nolen could not avoid noticing how her breasts swayed within her silk blouse, as she sat next to him. "Lately, I haven't had much time to enjoy any view," he answered.

She smiled. "If you would like, you may rest here tonight. You're going to have to refresh yourself once in awhile, between now and when we announce the gold strike. We need to ensure you stay healthy."

"You're right. If I'm able to get half of my normal amount of sleep, during the coming weeks, I'll be doing well."

"Then, before we talk shop, let's take care of you." She stood and walked toward the kitchen. "Are you hungry? I bought some steaks."

"Hey, thanks. I haven't eaten since breakfast."

Nolen eyed her feminine movements, then shook himself. You damn idiot, he thought. This is not the time to play around. Control your feelings. Be distant. Concentrate on business, and nothing else.

"Nolen, why don't you shower in the upstairs bathroom, while I finish cooking."

"Do I smell a little?"

"You smell just fine."

"Much like a bear, I suppose?"

She chuckled. "Go ahead; take your time. I started the charcoal, just before you arrived. It will be awhile before our meal is ready."

Within minutes, Nolen had adjusted the shower nozzle so surges of water pounded his tense muscles. Gradually, he accepted the fact that he had been moving too fast, and been too careful, for anyone to track him. He realized he actually felt somewhat secure, when he did not jump upon hearing a knock at the bathroom door, followed by a muffled question.

"What did you say?" he shouted in response.

Maida poked her head into the steamy room. "Sorry to bother you. How do you like your sirloin cooked?" She admired the blurred form of his body moving behind the fogged shower door, enjoying the opaque view.

"Pink inside."

"OK." She started to leave, then turned back into the room to collect his soiled shirt and pants. "I'll throw your clothes into the washer. They should be ready by the time we finish dinner." She slid open a closet door, selected a pair of terrycloth shorts and a robe, then placed them on the sink counter. "I'll lay out a comfortable change for you."

Nolen relished using the hot water to help soothe the pain in his lower back. It had constantly gnawed at him, since the flash flood. After showering, he massaged his tired muscles while toweling dry. By the time he returned downstairs, he felt refreshed.

Maida was placing large stuffed pillows around a glass coffee table in the semi-dark front room. She glanced up. "Your timing is perfect. I just removed the meat from the grill. Would you light the fireplace? My condo cools quickly, once the sun drops below the hilltops."

"Sure. I love an open fire. Flickering flames make me feel comfortable." He filled the sooty hearth with paper, kindling, and logs. The seasoned wood rapidly ignited, filling the room with warmth. He joined Maida at the table.

"I'll say this, you're a great hostess."

"It isn't all kindness. I was taught in business school— keep them happy, and they'll agree to almost anything."

"What do you need me to agree to?"

She hesitated. "In reviewing your plan today, I realized I am going to be the only one who remains fixed at their present day-to-day job."

"That concerns you?"

"Yes. If people are chasing you, they might find me instead."

Nolen smiled. "It's satisfying to know you are astute enough to foresee possible dangers. But you have little need for concern. I particularly want to ensure your safety. So, I've taken considerable time verifying that no traceable links exist between you and me. There's no paper trail or activity which CMR or the police can uncover, that could possibly connect us. For your added protection, I'm planning to hire a couple of bodyguards. They'll arrive in a week or so."

"I'll tolerate them so long as they remain inconspicuous. I want to minimize the disruptions in my life."

"Pretty lady, I'm sorry, but I disrupted your life the moment I offered you a partnership in the Association. Your life will never be the same again. If you had turned down this opportunity, for years you would have continually questioned that decision. Was I right? Was I wrong?"

"And now, that I have agreed to take that chance, I'm already asking myself those same questions."

"Eventually, you'll have the answers to tell your future children. No matter how this adventure ends, they'll respect you for taking the risks."

Maida quietly considered the insight of his comments. "If you trust me enough to be your partner in the gold mine, why did you stop telephoning last May, after I helped Digger with his taxes?"

"In the spring, my wife accused me of having an affair with every gal I spoke with. Since you're a beautiful woman, she grew insecure over my contacts with you, to dive for gold, or for any other reason."

"Were you having an affair?" The blunt manner in which she posed the question surprised Nolen.

"Well, no; however, under the circumstances, the thought did cross my mind. Things had been deteriorating in our marriage for some time. I think her jealousy was just an expression of her fear about us separating."

"Do you have any children?"

"Luckily, no. When we married, we agreed to allow our relationship to grow before we had kids, so she remained on the pill. The marriage lasted four years. Both of us matured and changed and recognized traits in each other. Many, we didn't like. We separated for the first time about two years ago, then reconciled. More recently, we decided to separate again. When we complete our settlement, we'll be legally divorced."

"You seem to still care for her?"

"In some ways, I always will."

"What caused the breakup?"

"Divorce stories always have two sides. If you spoke with her, I'm sure you'd learn that I have faults. From my point of view, several things were intolerable. First, I simply couldn't afford her. She would run up atrocious bills. Even with both of us employed, I could never balance the budget, and a savings account was out of the question. Besides, she maintained a shield of armor around her heart, something I was unable to penetrate. She even preferred sex without getting wet."

Maida laughed. "You're kidding?"

"I'm exaggerating—somewhat. She certainly would not allow herself to enjoy the pleasures of her body. Let's just say, her needs and mine were not compatible." Nolen reached for his fork. "Look, I won't have anything to eat, if I do all the talking. It's your turn. Tell me how you became a stock broker?"

"I was raised on a backwoods Georgia farm, near a little place named Adel. 'Adel, so close to Hell, you can see Sparks.' That's a jingle we exchanged in high school, to kid each other about the tiny towns from which we came. I was so ashamed to be seen in church, continually wearing worn, hand-me-down dresses. When I was fourteen, I promised myself I would not be poor all my life. And in school, I proved I was as capable as any other student. High grades allowed me into college. But I had to work two jobs to supplement what little my family could contribute toward my tuition."

"Sounds as if those were tough years for you."

"I recognized it would require hard work to succeed. Business administration and economics were my major. During my senior year, I decided to enter the brokerage business. It was a wise decision. In the last few years, I've grown into a successful businesswoman. I'm proud of all my accomplishments."

"I'm surprised you didn't label yourself, a 'business person'."

"Oh no. I'm competitive, but I don't require such phrases to establish my equality with males. I'm satisfied with my professional skills and with my femininity. I'm very comfortable being a woman."

"I noticed." Nolen shifted his gaze, away from her cleavage.

An hour passed, as Maida explained how her cousin had invited her to San Francisco, seven years earlier. She related her anxiety, searching for a job in the unfamiliar, overwhelming city. And later, how proudly she bragged over her first sale to her father. She recounted the long hours spent analyzing investments, endeavoring to provide the best possible advice to her clients. Her diligent efforts succeeded on several large investments, establishing her reputation, and generating referrals from satisfied customers.

Nolen threw another log on the fire, while Maida stepped into the bathroom to apply fresh lipstick and lip liner. When she returned, she suggested, "Why don't we spread the pillows near the fireplace? It will be cozy and relaxing, lying on the rug over there."

They settled side by side, enjoying the soothing warmth and the peaceful stillness of the evening. Maida asked, "When did you begin prospecting?"

"Digger taught me how to pan my first shovel full of black sand, when I was twelve. All through high school, I kept dreaming I would discover a nugget, two feet thick. Searching for gold is addictive; like a drug, it steals your soul. It's ironic that Digger found a huge strike, but hasn't had the opportunity to enjoy his success!"

"Nolen, when we first met, you told me how you needed to control your own lust for gold. It seems, that it has finally captured you."

"No, though I'm sure it appears that way. I gave Digger my word that I would take care of Hilda. All my efforts are for them. My addiction isn't to get rich!"

"What is your addiction?"

"Having a relationship, like Hilda and Digger's. They've spent their entire lives on the hunt, rarely finding enough gold to cover the rent. But they share a love for each other that is more precious than all the gold we've found."

Nolen's robe draped open as he spoke, revealing his chest. Maida noticed the soft, curly blonde hair covering his strong body. Her gaze drifted down to his hips.

"Thinking about something?"

She quickly sat up to light a nearby scented candle, attempting to hide how distracted she had become. "Yes...ah...sorry. So, what happened to you after high school?"

"I figured there were better ways, than prospecting, to make a living. Ways, not requiring a hand-to-mouth existence, worrying each week whether there would be enough money to pay the bills. After junior college, I joined the Army. Luckily, I earned a direct commission as an Infantry officer. Spent some great years traveling around the world.

But I returned home when the doctors diagnosed Hilda with cancer."

He moved a stray strand of hair from Maida's cheek, and draped it behind her ear. "My biological parents were Italian. I have always been attracted to women with thick, black hair." He hesitated, stiffened, and retracted his hand. "Sorry. I didn't mean to intrude on you. Sometimes, I unconsciously touch people I care about. Once in awhile, that gets me into trouble." He moved his body away from Maida, and stared pensively at the fire.

"It's all right. I don't mind."

Since he came downstairs from the shower, they had chatted constantly. Now, a silent anticipation filled the room. Nolen turned back toward her. Their eyes met and lingered. She leaned forward to kiss his lips. Her time had come to enjoy what she had fantasized about for months.

She slipped her hand around his neck and caressed his curly locks, urging him closer to her. Their tongues played, darting in and out of each other's mouth. Nolen paused, enjoying her willing tugs, then gently turned her chin, and began licking her ear. As his tongue tickled the side of her throat, she moaned and shivered. He gently touched her, his hand slowly wandering up and down the front of her body. He smiled, watching goose bumps surface on her arms.

Nolen stopped, hesitated, then placed his hands on her shoulders and gently pushed. "Stop. This is foolish. Our lives are complicated enough just trying to orchestrate a gold rush. You are sensual, and very inviting. However now is not the right time to become sexually involved."

Maida abruptly stood. She walked to the balcony door and snatched a sweater from a wall hook. She fumbled with the garment, attempting to drape it over her shoulders. In frustration, she threw it onto the floor. She sighed, then slid the glass door aside and slipped into the cool night breeze.

Nolen rolled onto his back and took a deep breath. He tried focusing his attention on the crackling sounds of the burning logs. Perspiration trickled from his armpits. He turned and looked onto the dark balcony. City lights, shimmering in the valley below, silhouetted Maida's lovely figure. He knew she was staring at him. Now what, he thought? Remember, you're supposed to prevent anything from touching you!

Seconds later, he made his decision. He rose, retrieved her sweater, and joined her on the redwood deck. Maida turned her back to him, leaning against the railing. He laid the wrap over her shoulders. The scent of her body near him triggered his adrenaline. He hugged her. She resisted, only until she felt him lick the nape of her neck, using the tip of his tongue to slowly trace small circles across her skin.

She arched upward against his chest, then turned around, embracing him, kissing him. Penetrating the opening in his robe, she fondled his muscles with her hands. Teasing, she bit his chin, kissed his throat, then began nibbling one of his nipples. As she tickled his chest, she whispered, "Since we both agree to quit resisting, let's step inside."

Hand in hand, Maida led Nolen through the darkened house, to her upstairs bedroom. A night light bounced a pastel glow off the pink walls of her boudoir. She faced him and removed a barrette from her coiled hair. Shaking her head, she set her luxurious curls free. Slowly, quite deliberately, her delicate hands unbuttoned her blouse, elevating Nolen's excitement as he watched her. She stepped into his embrace.

With simultaneous eagerness and hesitation, Maida and Nolen stood beside the king-sized bed, appreciating the thrill and anticipation of making love with a new partner. They delighted each other by sampling private recesses, which moments before had been forbidden. Both wondered what sexual boundaries they would soon cross.

Nolen bit her lower lip and sniffed her warm fragrance. When he slipped his hands inside her bra, she eagerly rubbed her broad breasts against him. His thumbs circled over and around her protruding nipples. She enjoyed her shudders, each time he touched her swollen nubs. He paused, waiting for any sign of reluctance, but met only her encouragement.

Maida slithered her hands down Nolen's sides to the front of his robe, untying the knot. She yearned to be tantalized to higher and higher levels of passion. It had been too long since she had chosen a lover, who delivered the stamina she desired, and whom she also wanted to spend time with outside her bedroom. Now, it was time to determine whether Nolen would meet her expectations. She slipped his shorts below his lean hips.

Nude, among the satin pillows and plush comforter, they shared their passion. Taking turns, they discovered each other—a pressure that excited, a lick that was unbearable, a position they favored. They played for half an hour with their mouths and fingertips, happening upon one another's secret pleasures. Maida was pleased with Nolen's gentle touch, openly giving, not merely taking. Finally, she grooved her sharp fingernails down Nolen's back, demanding his hips closer to hers. In time, with her insistence, Nolen drove his body against Maida, all the while whispering sexual fantasies into her ear. She responded with internal explosions. "Now! Now! Now!" she urged with delight.

Following her third orgasm, Maida relaxed her body clasp from Nolen and released a deep sigh. "You're very satisfying," she whispered, expecting Nolen to pull out of her, thereby letting her excitement fade. Instead, he rolled onto his back, drawing her on top of him.

"I'll never leave you unsatisfied," he promised.

She smiled. "I'm fine. You don't need to do more."

He began slowly pumping, while tantalizing her with his thumb. She gasped, when another series of internal shudders overtook her.

Finally, she softly slapped his chest. "No one has ever kept me cumming like that. Be kind to me. If we don't stop, tomorrow I won't be able to walk, or see."

They lay like two matched spoons, his shoulders and buttocks snugly touching her breasts and abdomen. She petted him, stroking his taut stomach, kissing his neck, and luxuriating in the pleasure of being appreciated and satiated. "I was right."

"About what?" he wondered.

"When you are around, my life does become more exciting."

Nolen grinned. "I'm glad to know I'm not boring."

They rested in their soft embrace. He dozed for a brief time, until his nap was interrupted by the sound of her voice.

"May I ask you a personal question?"

"You can ask any question you want."

"I noticed, while we were making love, that you have, oh it must be, thirty short, white scars scattered over the length of your legs. As though you were cut, over and over again. How were you hurt?" Maida felt a sudden tenseness crawl through Nolen's body. She waited for his answer.

"I'll explain, but I won't dwell on the subject. When I was eight, my dad remarried. He made a bad choice; at least it was for me. Shortly after the wedding, his new wife began abusing me whenever he traveled on business trips. Her favorite weapon was a three-foot-long, narrow, aluminum rod. Light and flexible, it never broke. At least, not until I chopped it up with an ax! It hurt like Hell when she struck me. The sharp edges cut my skin, causing those scars."

"You mean your stepmother beat you, often?"

"She was no mother to me."

"In my family, we may be poor, but there is an abundance of love for all the children. It's difficult for me to comprehend how someone could ever harm a child." Maida hugged him. "I'm so sorry it happened."

"Don't worry. When I was a teenager, I worked through the guilt and doubts caused by the abuse. My self-image is great. I wasn't a bad kid. My dad's wife was the sick one. Anyway, there was a bright side to the whole affair."

"What possible good could there be?"

"I learned that every person has the right to be respected as a unique individual, with something positive to offer to others. It doesn't matter whether they are strong or crippled, black or white, male or female."

"It's surprising that you were able to extract something positive, from such a dreadful experience."

"I wouldn't volunteer to go through those events again. But that difficult time in my life honed me like a fine steel sword, very hard on the surface, yet flexible."

"I'm glad, tonight, you allowed me past your hard surface," Maida admitted.

"I became stronger each time I withstood the loneliness, the battering at home, and the ridicule I received from my schoolmates, regarding my bruises and cuts."

"You don't sound bitter. I think I would be."

"Rather than bitterness, I developed an inner strength. I learned how to survive. Whenever I'm knocked down, I immediately stand up. If I can't stand, I get to my knees. I never stop fighting for my freedom. I never give up my humanity. It appears to be a physical thing. Actually, it's all mental. That honed strength is an asset. There have been many times in my life that I have tapped that reserve for my protection."

"Is that the reason why a coldness overtakes you on occasion? It's difficult to explain, an aloneness, almost a brutalness."

"Learning how to survive left a cold spot within me. It's an inner anger. One that I have never been able to eliminate. It must be part of the animal that resides within every human. That section of our souls from which violence erupts. A part some people willingly let control them. Just like dad's second wife. It's my ugly side. The side I constantly attempt to master."

Maida teased him. "What other dark episodes in your past should I know about?"

Nolen rolled over to kiss her. "Pretty lady, I've just told you more about my life, than I've ever shared with anyone else."

"Even your ex-wife?" she quizzed.

"Let's leave the past and focus on this moment. It may be a long time before we can play with each other again. A hundred things may block us from being together."

"You assume I'll want to jump into bed with you again."

Rather than answer, he cuddled her close to his pungent body and nuzzled her breasts.

"You're right," she whispered.

18

Tracking

19 August

Strands of midnight ocean fog drifted between the small wooden houses, crowded alongside the narrow city street. Pale shafts of light sliced into the darkness, from the front windows of two homes, illuminating the mist curling through the towering eucalyptus trees.

Boodan's driver increased the volume of the scanner radio, taped to the base of the dashboard of the rental car. For several hours, there had been the usual evening reports, broadcasting over the Monterey police net. Now, Boodan listened intently to the late evening instructions of the police dispatcher. With the aid of a pen light and a city map, he plotted the location of the disturbance, where the dispatcher had sent the officers. Satisfied that no police cars were near his vantage point, he lifted a starlight night vision telescope to his good eye. The device allowed him to clearly survey the neighborhood.

Boodan scrutinized Martin's dark, one-bedroom house. He detected a cat, crouched beside an alley fence, with its tail twitching. The feline crept forward, then paused between two garbage cans. Suddenly, it pounced. Boodan laughed when the cat lifted its head. A mouse dangled from the animal's jaws, writhing in a dance of death.

Boodan spoke into a hand-held radio. "Team one, get moving." A block away, a set of lights flashed on behind him.

A dented, rusty Cadillac slowly crept down the street with rock music blaring. Two men hung out the rear windows, laughing and hollering at one another.

Boodan turned to Luke Daniels, who sat in the rear seat. "Let's go," he ordered.

Both men, dressed in dark clothing and gloves, slipped down a side alley. They stopped when they reached the wooden fence, which enclosed the back yard of Nolen's house. Crouching against a stand of bushes, they waited until the car rolled past the front of the house. Then, each man climbed over the fence.

The car stopped. One passenger clambered onto the sidewalk. He walked toward a house across the street from the target building. Dogs began barking, alarmed by the noisy men and their strange car.

"Hombre, it no look like the house," yelled the second passenger, still hanging out the rear window.

"Si, this be it," the farm laborer on the sidewalk replied.

"No way, bro'! She don't live here."

The driver yelled, "Come on, amigo. Let's go."

Several neighbors peered through their windows, wondering about the unusual commotion in the street. The Mexicans noisily continued down the avenue, pleased with the one hundred dollars they had so easily earned.

In the rear of Nolen's house, Boodan picked the lock on a patio door, leading into the garage. They entered and kneeled in the dark, listening, while a dog next door continued to bark. Boodan keyed his hand radio. "We're in."

"OK," Boodan's driver answered.

The door into the kitchen of the home rendered some resistance before it opened. The burglars stiffened, when they heard the next door neighbor yell at his dog to stop barking. Expectantly, they waited a long minute before searching the rooms.

Each man carried a small bag and used a pen light to illuminate tables, shelves, and closets. They searched for clues to Martin's whereabouts, anxious for some means of pressuring him to forfeit his strike to Tower.

Boodan invaded the den office. Computer and mining books lined a bookcase. Scattered papers covered a roll-top desk. He took an address book, and a letter mentioning Nolen's pending divorce, together with a picture of Nolen and Digger.

Meanwhile, in the bedroom closet, Luke located a plastic file box. Leafing through the contents, he found a copy of a will, which he promptly stuffed into his cloth bag.

They had been rummaging for almost fifteen minutes, when their radios squealed. "Get out," Boodan's driver ordered. "Cop cars coming." Each man acknowledged the message and scrambled for the exit. As they crashed their way through the garage, the dog next door resumed barking. Both men tossed aside their radios, gloves, and lock-picking tools. They sprinted to the fence. Once in the alley, they ran to the next street. From there, they continued, choosing different directions. Luke rushed to a small park. He stuffed his bag underneath a previously selected bush, before hurrying toward the ocean.

Boodan hopped into his car, as it pulled away from the curb. It turned down a side street, just as a police cruiser stopped two houses short of Nolen's residence. A block away, another police car scoured the neighborhood. One policeman spotted Luke, with a torn pant leg. Reluctantly, the big man halted when the policeman switched on his flashing lights. The officer grew increasingly suspicious, while Luke explained his reason for a late evening stroll. When the stranger was unable to provide any identification, the patrolman elected to take Luke to the station, for further questioning concerning the reported burglary.

20 August

As Cholo Cantera followed the coastal highway into Monterey, it seemed to him that Mother Nature was struggling to decide what the morning weather should be. A half-mile off shore, a crescent-shaped fog bank stretched a gray shroud around the harbor. One end of the fog bank anchored south of the hamlet, sending mist tendrils crawling up the ridge. Three miles away, the other end enveloped the motels stationed along the coast highway, while from the east, bright sunlight flashed out of a blue sky. A warm inland wind pushed against the ocean fog, burning it back from the green hills and white houses crowding the rocky shoreline. Light and dark, warmth and chill, jostled each other, eager to possess the town.

The FBI agent drove inland, winding his way to the county courthouse, with its adjoining police station. After the district attorney verified his credentials, Cantera was assigned a police sergeant to assist him.

The policeman smiled. "What can I do for you?"

"We're investigating an individual, suspected of being involved in the theft of ammunition and weapons from Hunter Liggett."

"Well, we should be able to help. Army CID sends us reports regarding all felony incidences which occur on military property in the county."

"The suspect is named Nolen Martin."

"Who did you say?"

"Nolen Martin. His local address is..."

"210 Nueva Drive, right?"

Cantera nodded. "How do you know about him?"

"Last night at about 12:50 a.m., his residence was burglarized. We nabbed one fellow, two blocks from the house, who may have participated in the crime. I just heard about it during the morning briefing. The suspect is in the holding cell. Name's Luke Daniels. Caucasian. A real heavyweight.

Six-feet tall and solid. His face appears frozen in a permanent scowl. Reminds me of an angry bear."

"Are you sure this guy is involved?"

"He claims he's on vacation and accidentally left his wallet in his room. However, he just happens to have a California arrest record for assault, fraud, and car theft. He was picked up four miles from his motel. His pants were torn, and he wasn't acting much like a tourist. That was enough to hold him on suspicion of burglary. Our forensic team is searching the house this morning. Right now, we're waiting until we uncover some additional proof. If his prints don't show up, we'll have to release him."

Cantera jotted in his notebook, trying to concentrate and not appear surprised. "Can I gain access into the house this morning?"

"Sure, I'll tell Dispatch to notify the team that you're headed out there."

Cantera followed the sergeant's directions, leading him through the shops and businesses hugging the choppy bay. Eventually, he entered the more spacious residential areas, populating the southwestern end of the peninsula. Three blocks uphill from a verdant, green golf course, he located the single-story house. It was outlined with yellow crime scene tape, strung along the wire mesh fence, protecting Martin's diminutive front yard.

Cantera checked the garage. None of Martin's registered vehicles were there. Inside the house, Cantera worked his way through the rooms, which the investigation team had previously examined. Cantera carefully inspected the den desk, bookshelves, and indecipherable computer printouts. He was unable to find any evidence of white supremacist propaganda. There were no political documents of any type. All the books related to mining precious minerals. A bank statement, spread on the desk, provided Nolen's checking and savings account numbers.

While the investigation team worked the last room, Cantera decided to interview the neighbor who had reported the burglary. After Cantera rang the doorbell of the nearby bungalow, the front door snapped open, presenting a stubble-bearded man. "Yeah, what do you want?" he grumbled.

"Sir, I understand you reported the burglary."

"Yes, I did. I explained that to the policeman last night, and then again this morning. You'd think all you guys could talk to each other. I have lots of things I need to do."

Cantera smiled, showing the man his gold badge. "I did talk to them, sir. And I'm sure you're busy. However I'm from the FBI, and I want to personally hear your story. Your input could be the key to solving this case."

"Really? The Federal Bureau of Investigation." The man's face beamed. He turned and shouted into the house. "Margaret, come here. There's an FBI agent here who wants to talk to us." The man turned back to Cantera. "Just want my wife to meet you."

"My pleasure, sir."

"FBI, huh. What's Martin done? He always has been a bit different."

"Different?"

"Well, nothing bad, of course. Real independent, you know, stubborn-like. I guess that war in Iraq got to him."

As the man continued offering information, Cantera took shorthand notes on a small pad. "Has he done anything odd lately?"

"Got divorced. All them long Army separations made Martin's wife leave him. Anyway, that's what his wife said. But he's been a decent neighbor. Not very talkative though. My wife claims it's 'cause he's a prospector, chasing dreams." The man peered over his shoulder to see if his wife was nearby, then whispered. "He's been prospecting, off and on, lately. Hasn't found nothing."

"Have other problems been bothering him?"

"Last month, the damn prospecting landed his father in the hospital. My wife told me Hilda and her boy saw Digger slip into a coma. God bless him! Must have been tough on the old woman. There was...something...sort of strange. Nolen called me two days ago. Asked me to watch his place, real close, for the next few weeks. Like he knew someone would come snooping around."

A plump woman, wearing a flower print dress, huffed along the hallway. Cantera smiled at the lady and extended his hand to shake hers. "Officer," the woman asked, "would you like to come inside and sit? I have some fresh coffee on the stove."

He accepted her offer, following the couple into their sunny rear kitchen with its view of the ragged lawn. Seated at the kitchen table, the three chatted for several moments, as the agent gradually maneuvered the conversation back to the burglary. "Sir, you stated Martin telephoned you, two nights ago?"

"Yes. He said he was up north and would be gone longer than expected. Well, when my dog, Scooter, started barkin' in the back yard last night, I went outside to look around."

"We leave our dog out there at night," the wife interjected.

"I'd expected Scooter to be yappin' at those damn creeps driving down the street," the man continued. "But he was a pawin' at the fence next to Martin's garage, ignoring them Mexi...ah, guys in the car. So, I jerked him away and listened. After awhile, I heard me some unusual noises, coming from Martin's place."

The woman interrupted. "That's when I called the police. My husband was real brave. He stayed on the porch and waited."

"I turned off the back porch light and peeked over the fence. If you stand on your toes, you can see into Nolen's patio. Just before the cops got here, two guys ran out of the garage, likity-split. Shocked the heck out of me. They jumped the fence into the alley. I would've had a heart attack if they'd came into our yard."

"Did you see if they were carrying anything?" Cantera prodded.

"No, I ducked when the door banged open. I bet they was as scared as me, 'cause it sounded like one of them tripped over a garbage can, after he climbed that fence. Funny thing, how'd them guys know the cops were a comin'? The police didn't use a siren, not even once."

"Did Mr. Martin indicate where, up north, he was calling from?"

"He told me he was near Susanville."

Cantera thanked the couple for their assistance, then returned to Martin's residence. Checking with the investigation team leader, he learned what little they had discovered. "Well, the crooks appear to be professionals, almost too professional."

"How's that?"

"The thieves dumped all their tools as they escaped across the rear yard...radios, pen lights, and packets of lock picks. There were no fingerprints anywhere, not even on the batteries in their two-way transmitters. It's unusual for burglars to use radios. That indicates they had a man outside watching the house, while they worked inside. We've never had anything like this happen around here. Hell, there's not that much of real value to steal on this side of the peninsula. An M.O. like that might be used to hit a rich mansion down the road in Pebble Beach. But here? No way."

"Did you find any blood, or any cloth on the fence?"

"We spent thirty minutes searching there. Didn't find anything we could compare against the suspect's trousers."

"Any literature on political activities? Or perhaps, an address book left in the house?"

"Nope, nothing like that."

"May I borrow a computer sheet from the den? I want Martin's employer to explain it to me."

"Sure. Just return it to us, before you leave town."

Later, at the police station, Cantera reviewed the file on the burglary suspect. He learned that Daniels had rented room 15 at a local motel, the day prior to his arrest. A phone call to the motel desk clerk verified that several other men, who had checked in with the suspect, unexpectedly departed shortly after 1:00 a.m. The clerk also described the location of the suspect's room. He explained that only one person was listed on the registration card, and that only one pass key had been issued.

Cantera discussed, with the sergeant assigned to support him, a scheme to obtain more information. The young FBI agent received assurances that the Desk Sergeant would delay Daniel's departure, until the local investigation team returned. But without further evidence, the suspect would have to be released soon.

Cantera left the courthouse and hurried to a hardware store. After purchasing a glass cutter, file, and tape, he drove to the motel, parking one block away. From a phone booth, he telephoned room 15. When no one answered, he walked to the rear of the motel.

A retaining wall formed a narrow alleyway, parallel to the rear of the structure. While trying to avoid tripping over broken boxes and debris, the agent padded down the corridor. Carefully, he counted the number of windows he passed, until he reached the rear of the suspect's room.

Cantera pressed his ear against the window, listening for any sound inside. Detecting no noise, he began cutting a small hole in the corner of the window. Two minutes elapsed, then he pressed a piece of electrical tape against the glass and began tapping along the scribe lines.

A small triangle popped from the pane. He filed the edges of the glass triangle, then Scotch-taped it back into its hole. Twice, he verified that he could, smoothly and quietly, slip the wedge of glass in and out of the window. He stepped back to admire his work, satisfied that the suspect would not become suspicious of the unobtrusive crack in the corner of the window. This basic eavesdropping technique would have to do, he thought. Unfortunately, he did not have enough evidence to influence a judge to approve a wiretap.

Later that morning, the FBI agent trailed Luke Daniel's taxi, from the jail to the motel. Cantera moved to the rear of the building. As he approached the window, he heard the shower splashing. Cantera waited until the bathroom sounds subsided, then cautiously tugged the glass plug from its corner.

Thirty minutes passed. Cantera began to wonder whether or not he would learn anything worthwhile from his "brilliant" maneuver. Finally, he heard Luke speaking. The FBI agent leaned his ear closer to the hole, while slipping his note pad from his jacket.

"I want to place a long distance call to Nevada. Area code 702-899-5685. This is Luke. Yeah, got released about two hours ago. They had no proof. Did you find my black bag? Great. How'd the cops know we were in that house? OK, I'll hang around for two more days, then get back there. Right, I'll check in first thing, when I reach town." Luke dialed again. "Is this Oriental Massage?" he inquired.

The agent re-secured the piece of glass into its original position and stepped down the walkway. Pleased with his morning's accomplishments, he thought, Well, well, well, perhaps I am on to something now.

Upon returning to the police station, Cantera explained he had not learned any information which would prove the suspect guilty of burglary. However, he planned to tail the suspect for another few days.

22 August

Cantera waited, as the phone buzzed, followed by the standard words. "Recorder 18. State your message after the tone. Beep..."

"Update report, number three. I'm at the Monterey airport, following Luke Daniels. He is the only lead I have on the group that broke into Martin's house. I believe Daniels participated in the burglary, based upon the call I overheard him make to Reno, Nevada. He has purchased a ticket to Los Angeles International Airport on the nine o'clock flight on Air West airlines.

"I spoke with the computer firm owner and several of Martin's co-workers. The owner informed me that Martin had requested three weeks vacation, to assist his sick mother with her comatose husband. Martin's supervisor also analyzed the printouts I found in Martin's home. He said they were probability charts, of some type."

Cantera paused, flipped a page in his notebook, then continued. "Martin's foster mother maintained he is fulfilling a pledge to her husband, to search for a gold mine near Susanville, California. Coordinates listed on the computer sheets seem to confirm her statements, since they correspond to the region northeast of Susanville.

"Mrs. Jorgenson claimed she spoke with Martin one day prior to the burglary. He told her he was in Susanville and would be prospecting for several weeks. That same day, he also phoned his neighbor, again saying he was in Susanville. He asked the neighbor to watch his home. However, while checking Martin's recent bank account transactions, I learned that on the same day he spoke with his foster mother and with his neighbor, he withdrew $4,000 from an Oakland branch bank.

"In summary, Martin was in Oakland—not in Susanville—needed the use of his funds, and felt someone might be observing his residence. For unknown reasons, he has been lying to people about his activities and whereabouts. His behavior leads me to believe he is hiding from someone."

Cantera checked the small airport lobby, to ensure Luke was still seated, engrossed in a *Penthouse* magazine. Satisfied that the man had not moved from his chair, Cantera continued with his message. "I also researched Martin's telephone bills for the period, January through August. Two numbers were repeatedly dialed. They may furnish us additional information about his present activity and location.

"The first number belongs to a local attorney, who has handled Martin's legal matters for several years. He is presently representing Martin in his divorce. The lawyer said Martin is stable, but he is experiencing the normal financial difficulties accompanying a divorce. The attorney also indicated Martin seemed to be coping with his latest personal problems quite well.

"The second number was to the Virginia residence of the parents of his ex-wife. I suggest Bureau personnel, in Virginia, question his former wife regarding the suspect's recent activities, mental state, etcetera."

Cantera spotted Luke walking toward the boarding gate. "When my suspect terminates his travels, I will submit another report."

19

Credit

23 August

In a rented cabin near Susanville, Nolen finished trimming his new beard and looked out of the cramped bathroom. Maida lay in bed in front of the gray stone fireplace. He smiled when he realized she had fallen asleep, since he reluctantly disengaged himself from her sensual embrace. Toweling his hair dry, he stepped to the edge of the bed, leaned down, and kissed the back of her exposed neck. She stirred.

"I'm tired...let me rest," she mumbled.

"Time to wake up, pretty lady. I made a fresh pot of coffee and let you sleep an extra forty-five minutes. We have to get moving. The less time you spend near me, the safer you will be."

"You're cruel. I woke at 4:00 a.m., and drove two hours so I could see you. Now, after satisfying your desires, you're forcing me out of bed."

Nolen bit her ear, then slipped his hand under the thick down comforter to stroke her warm lower back. "I must have mistaken your moans earlier. I sure thought you were being satisfied."

"I confess. I faked an orgasm, just to make you feel macho. Now, go away."

"Your superb acting convinced me! Especially when the headboard cracked as you were pushing and pulling on it. You should go to Hollywood and become an actress."

"The only place I want to go, is back to sleep."

"Maida, even though I enjoy being with you, I don't want you taking any chances. So, while we are dressing, tell me what has been accomplished since we formed the Association. Then, you must leave."

Grudgingly, Maida pushed herself upright, donned one of Nolen's cotton shirts, yawned, and ran her hands through her dark curls. She accepted a steaming cup of coffee from Nolen. "Well, I finished the first of my assigned tasks by establishing a twenty-four hour answering service. It will record messages from all our scattered partners. I thought it would be wise, if we assign code names to each person. Then, have them leave only a telephone number and a contact time with the answering service. Later, I'll use a direct line to return their call. That way, no third party will ever overhear our business plans or progress reports."

"Sounds good. Make all your return calls from different pay phones, so we reduce the chance of someone tracing you."

"That's the procedure I have been following. I do not want to meet any CMR people."

"Maida, call me every day at 8:00 p.m., with an update on what our people have achieved during the previous day. You'll need to keep it short and to the point, in your usual efficient manner."

"Sure, just leave your number with the answering service by 7:30 each night."

"What about other communications, Maida?"

"The post office boxes have also been rented."

"Where's mine?"

"It's in a little place called Doyle, about fifteen miles west of here. Your mailbox is at the post office in the center of town. I rented it under an alias, of course. Here's your key."

"Super. Tell everyone to send in summary reports, whenever they achieve major progress on their tasks. You can consolidate them, and forward any documents you think I need to study. What else should I hear?"

"Well, before your brother began traveling around the West Coast, he hired two of his college classmates, who are lawyers, together with another attorney he had previously employed. One is an expert in corporate law. Another specializes in criminal law. The last is a business attorney. The business attorney should link up with you early tomorrow. He will help the local banker establish and manage all rental and supply contracts."

"And the estimated costs for the legal assistance?"

"Paul hired them to form the Law Department for the Association. They agreed to a flat fee of $9,000 a month for each lawyer, with bonuses should they achieve specific milestones prior to filing. Bonus payments could double their monthly pay to $18,000. Therefore, our near-term fixed monthly expenses are $54,000. After we file, and the claim is legally accepted, their total monthly fee should rise to $83,000."

"That arrangement," Nolen grumbled, "seems expensive and excessively generous to the lawyers."

"I know; however, I agree with Paul. These men will know every strength and weakness of the Association, every legal and not so legal twist and turn we make. When challenges and attacks threaten the Association, our lawyers will be accepting much of the heat, answering most of the questions. Providing them the opportunity to become millionaires gives them the incentive to work diligently to protect our interests, and tenaciously represent us through the difficult times."

"What about the Caribbean trip?"

"Paul and the corporate lawyer flew into the Cayman Islands yesterday, to establish the holding company. They will contact a local attorney in Georgetown to handle all our transactions, including depositing money into foreign bank accounts. The people there provide those services all of the time, so there should be few problems and even fewer questions. Tomorrow, the lawyer will continue on to the capital of Panama, to establish the land-purchasing company."

"I need to know the corporate names."

"The actual holding company is called Nippon International. The lawyer will incorporate another one, titled Martin Developments, through which we can continually filter money from minor purchases and sales. He felt, if the IRS does investigate, possibly they will waste their energies focusing on the fake holding company. The land purchasing firm will be named PanAm Properties."

"Great," Nolen said, "and what about bookkeeping support?"

"I titled it Package Accounting. Do you like that?"

"Cute." He scratched his itchy beard, then slipped on his suit.

Maida continued. "I've rented five offices; two in Susanville, two in Reno, and an audit office in Red Bluff, on the other side of the mountains in California. I allocated the business duties among the Susanville and Reno branches. Also, I've employed five supervisors and five CPAs. Twenty other accountants will be hired on a timed basis, as we need them."

"Excellent. Assign several of the best bookkeepers to the audit office. That way, all Association partners will feel assured their interests are being safeguarded. What do the costs look like?"

"When we're at full force, approximately $64,000 a month."

"That's not bad, considering your team accounts for millions of dollars worth of Association transactions."

"Speaking of transactions, I'd feel more comfortable, Nolen, if you would stay away from Susanville."

"Look, I'm not keen on bumping into any CMR thugs either. That's why I'm growing this beard as a disguise, and why I'm driving a rented truck. But today's deal is critical to the success of the Association. Hopefully, my recent research will pay off."

"Try not to be conspicuous," she pleaded.

"Thanks for caring, pretty lady. Now finish dressing. You need to be in San Francisco tonight."

At the doorway, they kissed, lingering momentarily, savoring the intimacy. They both understood it would be some time before they would touch each other again. Reluctantly, they broke their embrace.

Upon arriving in Susanville, Nolen twice cruised past an unimposing bank before parking one-half block away. When he felt satisfied that no one was following him, he strode to the bank entrance. Inside, he crossed the small, carpeted lobby leading to the executive offices. He stopped in front of the secretary's desk.

She smiled and surveyed the stranger dressed in a three-piece suit, which proved he was not a local customer. "May I help you?" she inquired.

"I want to speak with Mr. Hamilton Roberts regarding a large deposit."

"Let me verify that he's available, sir," she said, raising the telephone to her ear.

Moments later, she guided Nolen to a wood-paneled rear office, occupied by a middle-aged executive. He was a tall, unathletic gentleman. A double chin was beginning to sag on the man, and his stomach intimidated the buttons down the front of his white dress shirt. Nolen noted his toupee.

The banker stood to shake Nolen's hand. "Welcome," his deep voice boomed. "I'm Hamilton Roberts, Branch Manager. Please, sit down. How can I help you, Mr?..."

Nolen paused, until the secretary had closed the door, assuring their privacy. "The name's Martin. I'm considering depositing $1,600,000, composed of cash and another asset, into your bank. But only if you can provide me with appropriate services." Nolen snapped open his briefcase, revealing stacks of fresh green bills. "I have $149,000 with me today."

"That's a lot of money to be transporting unguarded."

"It's guarded." Nolen folded back the left side of his suit jacket, exposing a shoulder holster cradling his 357 Magnum pistol.

Hamilton's eyes widened. He moved his foot toward the police notification button under his desk. But he hesitated, realizing robbers walked out of—not into—banks, carrying thousands of dollars. Furthermore, the million-dollar-plus balance would make this newcomer one of the largest investors, within the ten branches of the tri-county banking system. Landing the deposit would be a personal coup, particularly since the bank's cash position had deteriorated, during the economic downturn that northern California had experienced in the past several years. As he reconsidered, he removed his foot from the alarm. "What type of assistance do you want?"

"In the next sixty days, I plan to open approximately sixteen businesses here in town. All their payroll, working accounts, and profits should be concentrated in one bank. Consequently, a single banker could provide special account handling. Of course, for that additional service, I expect to pay a handsome fee."

Hamilton recalled his wife's repeated, unkind accusations. She had yelled across the breakfast table, that it was his fault that the motel he had purchased at the edge of town was failing. She claimed his poor investments were devouring their savings at an alarming rate, including her inheritance provided by her father. She ignored his explanations regarding start-up costs coupled with a slumping economy. Her nagging retort was, he never intended to buy her a more prestigious house, as he had once promised. What hurt him the most was, when she declared her mother had warned her not to marry such a meek man.

Hamilton tried to sound casual as he asked, "How large a fee?"

"Twenty-five thousand a year. However, I would need my banker available, night or day, to receive calls from me or my partners, and to supervise speedy transfers of money, as we may direct. Can your institution accommodate my needs?"

"Absolutely," Hamilton eagerly responded. "To ensure you receive the best possible service, I'll assign two employees, specifically to handle your various accounts."

"Good, that's the type of support I'm seeking. One of my firms will be an investment company, for which I'll also require a financial advisor as Chief Executive. I anticipate this firm will generate over a million dollars in profit, during the first ninety days of operation."

Hamilton coughed, then replied, "I'm sure you have given considerable thought to this business venture, Mr. Martin. And I don't wish to dampen your resolve. However, I feel obligated to advise you that your profit estimates seem optimistic for the tri-county area which we serve. I hope you're not expecting much cash to flow out of our local economy. We enjoy very little tourist traffic, and most of the farmers and lumbermen around here are just barely surviving."

"You're correct, Mr. Hamilton. Our profits will be generated from people outside the county traveling here to spend their money. Since Susanville is centrally located to the site where most of my business will occur, I need a local agent who can assist my partners in arranging transactions. It's critical that I complete purchase agreements on time, so my team does not fall behind schedule."

"What type of agreements are you referring to?"

"Licenses, inspections, supplies, construction, plumbing, and electrical support. In fact, it would be most helpful if someone could aid my attorney in making our first major purchase, when he arrives tomorrow."

"What type of venture is this?" Hamilton was becoming more inquisitive.

"Now, I realize none of the good people in this town know me. Also, I expect most of the local merchants will not be used to moving as fast as I must. That's why I require the assistance of a person who is highly respected and trusted by the townspeople. Someone acquainted with the members of the city council and the local merchants. Someone to 'grease the skids', so I can rapidly get my companies into full operation. Do you know anyone qualified for this job?" Nolen watched the banker, hoping the dangled bait would produce the right response.

"I don't mean to brag, Mr. Martin, but I believe I can readily handle the Chief Executive position you have described. I've lived in this community over ten years. Furthermore, I own a lumber and hardware store, together with a motel on the western edge of town. All the important people in the local construction trades are personal friends. And recently, I was elected as one of the five City Councilmen and can be invaluable in that arena. Of course, there is my extensive banking experience."

"Well Hamilton, with your background, you recognize that jobs, sales, and tax revenues will significantly increase if a strong, new industry comes to Susanville. It would benefit all concerned, if you could influence the town government and local businessmen to appreciate the opportunity I bring to this community. Such support would materialize into immediate profits, both for them and for you."

"Oh yes, I can foresee the benefits, as I'm sure, will many other leaders in the community."

"Excellent," Nolen said. "Let me pose some specific questions. Could you help my organization establish operations inside Nevada?"

"Yes, I have several business and banking contacts in Reno."

"Can you find a man to supervise a security force?"

"Security force?"

"Right, a person experienced in coordinating guards and night watchmen. Someone who excels in that type of work."

Hamilton concentrated, scratching his cheek. "There is one person, Jake Samson. I could speak with him today, to determine his interest, and how soon he could be available."

"I've already explained your first two jobs, of representing our interests, and advising us on financial matters."

"And the other job?"

"Due to the magnitude of business I will bring to your bank, my Association requires a line of credit. We want six point six million dollars credit, which equals four times our deposit of $1,649,000."

"That is an exceptional amount of money." The banker swallowed, hoping his deal was not turning sour. "In these lean times, the bank's credit policy, for any investments over one million dollars, is to approve only two point four times the asset value." He entered the figures into a calculator as he spoke. "That's $3,957,600 dollars. Additionally, I would have to present your request to the other members of the Board of Directors. They would have some concern about authorizing such a large sum. Especially, for a venture which you have not yet fully described. After all, the Directors have had no previous experience with your management team."

"That is where your capabilities apply. You will be my representative. You will know what I'm doing."

Hamilton started probing, attempting to learn if Nolen was conning him. "Though your claims sound positive, we require substantial proof that your new businesses can repay this loan. The Directors would expect to know how the money would be used. In addition, they would demand controls over all withdrawals."

Nolen frowned. He had anticipated that the banker would reject his high credit request. The three point nine million dollar figure, however, would not provide the Association much leeway. All planned transactions would need to come in below the negotiated price levels that Nolen originally estimated would be agreed upon. But now was not the time to debate that issue.

"Fine, Hamilton. I can operate within the confines of that much credit. Let me caution you to exercise discretion, and minimize the amount of business information you share with the other directors or bank personnel. Any word concerning my business activities, or how I am conducting business, must not leak to the public during the next three months. To no one! To help prevent information leaks, I will authorize $55,000 in cash, to be stored in a safe deposit box. You alone may disperse those funds, to influence others to appreciate our need for privacy, to skip asking unnecessary questions, or reveal our business transactions."

The banker fiddled with his belt buckle. "That doesn't sound quite legal, almost like a bribe."

"Hamilton, view it as paying people to be our consultants. And you may use your own discretion with those funds. Do what you feel is best. Anyway, there is little to worry about, since you'll be using cash payments composed of small bills, making it difficult for anyone to prove any wrong doing."

"H'm, I see what you mean. Since I control the funds, I can ensure that all transactions remain legal."

"Absolutely," Nolen assured Hamilton.

"Also, as your advisor, I will be the only person with first-hand knowledge of your strategy and specific efforts. Though difficult, I could minimize critical business plan information from reaching the Board. I'm confident that I can satisfy your requirements."

Nolen smiled. "Well Hamilton, you're certainly the man I've been searching for to fill this position. The only item left to conclude is your signature on an agreement specifying your fee, and the scope of work you'll provide for my Association."

"Certainly. However, first I need to know more about your business plan, and then check your credit background."

Nolen let his expression turn stony hard and looked into Hamilton's eyes. "You must understand something. From this moment on, you will learn enough sensitive information to ruin the lives of the Association members. As long as you work with us, everything will be fine. If things become too difficult, you can even get out of our way. Again, everything is fine. However, you must sign an agreement to pay the Association a $100,000 penalty, if you leak the nature of our activities during the next ninety days, or if you attempt to block the Association's legal maneuvers."

The banker remained silent for a long moment. An ominous churning in his stomach made him fear his life would never be the same, if he agreed to Nolen's offer. Beads of sweat slipped down Hamilton's neck. *Does this stranger want to launder money from drug sales?*, he wondered. *If so, I'll cooperate for now, and notify the FBI as soon as possible.*

"Mr. Martin, you mentioned an asset other than cash. Will that be stocks, bonds, CD's?"

"Here is the asset." Nolen took a small nugget from his pocket and tossed it at the banker. The man caught it with a reflex action. "Gold, Hamilton—245 pounds of gold. It is the secret that I demand you guard."

Relief washed over the banker. "I was worried at first, thinking you might be trying to entice me into an illegal proposition." Hamilton wiped his brow with his handkerchief.

"On the contrary, your bank's exposure is low."

"Where did you obtain this gold?"

For thirty minutes, Nolen related the events surrounding the strike, the assay, and how he had formed the Association. "I'm offering you the same deal I gave my other three partners: four and one-half million dollars in ninety days, with an eight percent share of the mine revenues, plus a percentage of the special projects that you supervise. Will you join us?"

Hamilton considered the rarity of the lucrative opportunity afforded him. Certainly, this was the transaction that would make his wife proud of him. It might even satisfy her. "Well, there are great risks in this plan. Yet, it seems feasible. And the upside return is incredible, not to mention exciting. However, I will only sign the agreement after you deliver the 245 pounds of gold, and I can verify its authenticity."

Knowledge of the $149,000 deposit rippled through the staff of the small bank, as they watched the Branch Manager personally secure the bundles of cash, and open a new account for the bearded stranger. Curious eyes followed every movement of the heavy trunk during its limited journey from the mysterious depositor's truck into the bank vault.

Within an hour, Nolen sat in his rented truck. He felt nervous leaving all the Association cash and its trunk of gold with their newest partner. The signed personal service contract and the credit agreement provided little protection. Could Hamilton be trusted? Nolen wondered. Would he hide the gold, then cancel the credit needed to fuel the Association's scheme? Without appropriate financing our plan will wither and die! Nolen gagged, restraining his nausea.

20

Cinderella

24 August

When old man Tiller opened the red and white roadhouse nine years earlier, the few locals celebrating the event cautioned him that the name was too long for such a tiny bar and grill. The old man laughed, ignoring the warnings of his friends. He reminded them of the four-lane highway, proposed as a high-speed route across northern California to Reno. He elaborated how the highway would pass his doorway bringing numerous gamblers. They would not care about the length of the name, "Tiller Junction Lounge and Poker House".

Instead, the visitors would chance their luck with his slot machines, and when they traveled on, leave behind a large profit. Tiller claimed he was at the right place, at the right time. The old man boasted he would earn a fortune, just like all the owners of state line gambling houses situated along major highways in other parts of Nevada.

His friends wished him well, but cautioned him that he was the gambler. They were right. Six months later, state-wide tax reductions killed all hope of developing the sparsely-populated desert. So, old man Tiller lived his final years blasting off letters to state officials in Sacramento, declaring the narrow road dangerous, demanding that the two lanes be widened, and touting the merits of building a new highway.

The longer his pleas were ignored, the more bitter he became, while surviving on the few drinks he served to tired truck drivers and smelly deer hunters. The occasional tourists, who ventured into the lounge, quickly discarded their timid expectations that the roadhouse honestly advertised quality. The hamburgers, which were the staple of the bar, were either greasy or overcooked.

Now, the regulars glanced up from their late afternoon drinks, to survey the two strangers entering the tavern. A wrinkled woman alternately dropped nickels into the two slot machines remaining in the bar. She mumbled each time the tumblers did not favor her with the ring of a winner. In a side room, a young girl played with a kitten, frolicking back and forth across the pool table, batting a ping-pong ball over the worn green felt. The child giggled at the cat's scampering antics.

Nolen and Hamilton selected a table away from the bar, then ordered sandwiches and beer. They studied the waitress who had just taken their sandwich order. She was a thin woman with wrists favoring broom handles. Her eyes hung back in their sockets, two flashes surrounded by shadows, reflecting an age older than her years. A bandanna collected her long, gray-streaked hair into a sheath. Her hands were clean, but her fingernails revealed wear, cracked from long hours of preparing meals and washing dishes in the kitchen. She chose to ignore highlighting her underlying femininity with powder or lipstick.

A lanky truck driver, wearing a tan cowboy hat, called from the eight-stool bar. "Honey, how about coming over here and scooping me a big dish of ice cream?"

"Be right there, Buck." She frowned at the truck driver, before pinning the sandwich order on the clothespin carousel, then moved in front of the freezer.

"That must be her," Hamilton commented. "For a gal of only thirty-two years, she looks tired of people shouting at her."

Nolen answered, "You'd be whipped, too, if you lived here, working your fingers to the bone to maintain this place, while raising your kid."

The truck driver leaned over the bar. He pinched the waitress on her butt.

"Ouch!" she hollered, then instinctively spun around.

The driver rocked back on the rear legs of his stool, laughing at her displeasure and poorly-aimed jabs. "Now, Honey, you know you like my attention."

She calmed, then placed her hands on top of the wooden counter. "Buck, you are a real pain." She shoved his hands off the edge of the bar, sending him tumbling backward off his stool. The regulars roared with laughter.

"I'm sorry." The truck driver seemed surprised.

"Don't you sorry me. Just get up, and get out!"

"Seems as though she has some backbone," Nolen whispered.

"Yes," Hamilton replied. "We must keep that in mind. The county records didn't provide any hint as to her character."

"What did you learn?"

"She inherited the cafe/bar when her father died six months ago. Four thousand dollars in back taxes are still due on this small establishment. Court documents indicated her ex-husband has failed to pay his child support obligation of $125 a month. That's all the information I uncovered."

As the waitress served the sandwiches and beer, Nolen smiled at her. "Do you own this bar?"

She hesitated, eyeing the two strangers. "Why are you asking?"

"I'd like to discuss some business. Would you sit down?"

"Look pal, I've had a tough day. If you want someone to work on her back, your best bet is to go into Reno." She shoved their plates across the plastic tablecloth.

"I'm serious. I want to discuss buying your cafe. That is, if you're interested in selling," Nolen said, baiting her.

The woman paused, turned abruptly, and walked to the kitchen doorway. "Sam, watch the bar, will ya?" She returned to the table. "Who are you?" she questioned, as she sat down.

"My name is Nolen Martin. This is my partner, Mr. Roberts."

"You handled that trucker quite well," Hamilton commented.

"He's like a lot of guys," she grunted. "Always pressing their luck. Now, why do you want to buy this place? We're twenty-five miles from the nearest town, and rarely attract any tourists. Neither of you look like the types eager to live in a quiet, lonely desert."

"I need a tax shelter," Nolen casually responded. "Your bar would be an investment. I can write off the business expenses, wages, and improvements."

"Sounds hokey to me. How much are you offering?"

"Eighteen thousand for the building and the land. Five thousand for all the licenses."

"Mister, are you kidding me? That's nothing. The licenses alone are worth four times that much!"

Nolen knew his first bid was low. "Well, I'd say you're exaggerating somewhat. Many people are unemployed in this county. Anyway, I only want eighty-five percent of the place. You would still retain fifteen percent of the profits."

"Still not nearly enough. I don't think so."

The banker slid his chair back. "Nolen, let's try the other place?"

The woman raised one eyebrow.

"Wait a minute, Hamilton." Nolen leaned forward. "Miss, what about remaining as a part-time manager, plus receiving a salary. Say, $12,000 a year?"

"Well, that sounds better. But my daddy put a lot of years into this place. I couldn't sell for such a low price."

"How about a paid medical and dental plan for you and your children? You would also receive two years guaranteed employment, with two weeks annual paid vacation."

The waitress glanced at her little girl, still playing with her cat. "I'm just not sure."

Nolen turned to his banker. "Hamilton, maybe you were right." Both men stood up. "Thanks for your time."

"No need to leave before you finish your beer," she replied. "Tell me some more about your offer."

The two men sat down. Nolen resumed negotiating. Thirty minutes later, he again improved the bid. "This is the most I can offer. With this, I'm losing my shirt. You retain twenty-five percent of the business and receive $32,000 for the building and property. I pay $18,000 for the licenses, and you remain manager at $16,000 a year. I toss in medical coverage, plus four weeks vacation time. We split any back taxes. Agreed?"

"How do I know I can trust you?" she countered.

"My account is with First Mountain Bank in Susanville. You can verify my credit there. Any lawyer you select can review the contract. It's written in plain language. Hamilton, please show her the papers."

As she read the one page document, Nolen withdrew seven crisp one thousand dollar bills from an envelope, then laid them on the table. "Furthermore, I will hand over this good faith money to you now, after you sign the contract."

The waitress stared at the cash, fingering one of the fresh bills, purposely displayed in a fan pattern before her. "Excuse me." She went to the pay phone mounted on the wall near the cafe entrance. Following a brief conversation, she returned, and extended her hand. "You own a bar, mister."

"Great," Nolen said. "When do you want to meet with your attorney?"

"No need to waste any of my money on some attorney. I just called my sister-in-law. She's the head cashier at First Mountain. If your partner is the bank manager, I figure you're OK. Just fill in that contract right now with the amounts we agreed to. I want both your signatures on that paper, together with Sam's, as witness. Then, it's done!" She folded the bills, quickly stuffing them into her bra, anxious to change her daughter's life for the better, before the strangers changed their minds.

' "Hey, everybody," Nolen shouted, "drinks are on the house."

Later, when they left the bar, both Nolen and Hamilton experienced some difficulty in walking a straight path to their car.

"Hamilton, old buddy, I just thought of a name for her."

"Her, who?"

"I'll call her Cinderella," Nolen said turning to look one more time at the lounge. "Behold—our newest venture. She's a shabby, nondescript building now, but she'll grow into a spectacular casino. And while she's blossoming, we'll earn seven million dollars!"

24 August

A secretary tapped lightly against the office door of the senior FBI agent, then poked her head into his room. She waited for her boss to direct his attention away from the numerous papers spread across his desk. He scowled when she informed him Agent Cantera was calling from the Reno, Nevada FBI office.

The distinguished, elder Special-Agent-in-Charge turned to his private safe. He removed the red-tabbed Martin file, stamped EYES ONLY SAC. Grabbing the telephone, he barked, "What's going on, Cantera? I've had no report from you for several days."

"Sir, I've been following the burglary suspect. First, he flew into L.A., stopped for a day, then flew to Denver, stopped for another day, then caught a connecting flight to Reno. I trailed the guy to a duplex in Sparks, Nevada, where he met with five other men. I took digital photographs of each of them, and I will e-mail those pictures to you. After further investigation, I learned the group rented the duplex a week ago. Four phone lines have been installed."

"Was one of those numbers the same one the suspect called from the motel in Monterey?"

"Yes, and there have been other calls around the country. Later this morning, a supervisor at Pacific Bell will provide me with a complete listing of all numbers dialed from the duplex, during the last week. When I receive the printout, I'll do some additional follow-up, then forward those results to you. So far, all I know is that most of their calls have been directed to California and Texas."

"Son, you surprise me. I didn't expect someone so young to perform so well."

"Thank you, sir. Did we receive any response from the Virginia office?"

"Martin's wife did not provide any useful information."

"Sir, when I have the opportunity, I'll do some further checking with the military CID at Hunter Liggett, so I..."

The SAC interrupted. "Don't do that. Keep following this lead. It's hot."

"But..."

"Look Cantera, I'll get a federal judge to authorize electronic surveillance based upon conspiracy, burglary, interstate flight or whatever. I'll send you our best monitoring team. They will be there by 3:00 p.m. tomorrow. For now, stay in Reno. Ensure our team is safely positioned and begins their monitoring. Do you understand me?"

"Yes, sir."

"This evening, let me know the results of the phone activity for the suspect group."

The SAC hung up without waiting for Cantera's reply, then entered several notes into his file. He stared through the window at the Sierra Nevada mountains rising north of Sacramento. Absentmindedly, he flipped a pen around in his fingers, considering what to do next. The telephone message he had received earlier that morning still troubled him.

Washington D.C. had instructed him to contact the Sierra Army Depot to begin directing all investigations regarding attempted penetrations of the depot. Furthermore, he was to inform the Depot Commander and his security officer to communicate all future requests for assistance through his Sacramento FBI office.

To anyone overhearing the message, this action would appear routine. However, the SAC recognized a major decision had been made. The administration was distancing itself from any political repercussions, should information leak out about the secret research activities. If the SAC complied with the message, he would be the one held responsible for any operational failures.

Years earlier, everyone involved knew problems might arise. The Vice-President had concerns that the secret could not be protected long enough. Because of that risk, the key leaders hand-picked him for this assignment. As they had stated, "He has a track record of quietly and quickly handling delicate emergencies."

What was unstated, yet understood, was that they trusted him to sacrifice his career, as well as his family, without opening his mouth. His duty was to protect the administration at all costs. Ironically, the $500,000 personal contingency fund, prepositioned in his Swiss bank account, seemed like a paltry safety net, now that the threat of failure loomed more real.

The SAC picked up the phone to call Major Clement's office. Each time he punched a number, he realized he was placing himself into a win-lose situation, hero or traitor, all or nothing at all. If he were wise, if everything went well, if it was not already too late, they would succeed.

However, success would bring no fanfare. Instead, merely a quiet promotion, and a return trip to Washington D.C. Then, a message would be sent to the terrorists, who were working to plant biological bombs in large U.S. cities. He, together with a select group of men and women, would enjoy the satisfaction of knowing, they had out-maneuvered the radical Muslims who threatened America.

But if something went wrong, if they were found out, it would be impossible for him to prove he was acting on behalf of the FBI. The Bureau would deny any knowledge of his actions, or of an underground network of scientists secretly using government labs. The Bureau would declare him unstable, thus a threat to peace. He, and several others, would be sent to jail. The SAC quit hesitating. He stabbed the last number.

That night in Oakland, a businessman laid a nugget of gold on top of an assay report. He looked at Maida, as she adjusted her pinching earring. Having worked the bay area extensively for the past five years, Maida contacted several clients with adequate money and avarice to be inclined to accept her unique offer. This gentleman was fifth on her list. Just like the others, a gleam appeared in his eyes.

"Let me see if I fully understand all the facts about this investment," he said. "We're going to form a consortium to buy twenty percent of the voting stock of a mining company?"

Maida smiled, "Yes."

"All purchases will be completed prior to the announcement that this mining company has acquired a portion of the Association's new mine?"

"Correct."

"Since the stock presently sells in the eight dollar range, it should increase in value, by at least six times, to forty-eight dollars a share?"

"That's right," Maida confirmed.

"And I'm being offered a consortium partnership costing $80,000, which will grow six times to $240,000 in ninety days or less?"

"Right again."

"I authorize you to commit within the next twenty-four hours, $40,000 of my investment funds, to begin stock purchases, followed by another $40,000 purchase within two weeks?"

"Plus my commission," Maida reminded her investor.

"But, prior to the time of purchase, we must agree to hold the stock a minimum of forty-five days. And no matter what happens, the Association will guarantee a repurchase price of twice what we pay, in exchange for signing over special voting rights to you?"

"Yes, we simply want the ability to place our man in power as President of the company. That way, we ensure that we control the operations of the firm. You, of course, retain all other voting rights. Therefore, you can protect your interests in the company."

"Wait a minute," the man protested. "How will you prevent the SEC from learning you are coordinating a takeover?"

"The purchases will be made over a five-week period," she replied. "No more than two buys will be made from any one city, or by any single broker. I've already begun acquiring stock."

"And the Association gets paid ten per cent of my profit, only if the stock triples in price within the first ninety days?"

"Right again. The Association's profit materializes only after we deliver a huge return to the consortium partners. We felt such an arrangement would give each partner confidence that we have a strong incentive to ensure this investment evolves as described. We won't earn a cent until you are guaranteed a grand profit."

"You know, Maida, I always knew you were bright. I just never envisioned you would construct such an attractive offer."

"It helps to be privy to inside information," Maida responded, confidently.

The businessman grinned and stretched out his hand. "It's a deal!"

After pocketing the investor's first payment, Maida departed the office. She descended to the ground floor of the building and stepped onto the dark San Francisco sidewalk. Seated in her car, she attempted to relax, while re-confirming her schedule as well as her calculations, for the fourth time that evening. Nervously, she verified that she was not making any mistakes. Indeed, ten percent profit on the price rise of 200,000 consortium shares would produce $240,000.

I'm not advising any of the investors, she promised herself, that next week when I begin purchasing common stock for them, I will also start selling futures contracts. Speculators, in the futures market, will jump at the chance to acquire gold at twenty dollars an ounce less than they can buy it from anyone else. But once the strike is announced, the price of gold will drop eighty to one hundred twenty dollars below the cost of the futures contracts. Then, those same investors will pay to get out of their contracts, to cut their losses. Being conservative, if we average seventy-five dollars per ounce for 10,000 ounces, the futures contracts will generate another three quarters of a million dollars profit.

Since the mining stock is currently at six dollars a share, I can use a quarter of the investor's money to make judicious margin buys for the benefit of the Association. That risky move could net about $2,800,000 more dollars.

So the combined take, generated from a percentage of the consortium's stock profits, futures contracts, and margin buying, will be $3,790,000. An even greater fortune could be made if the Association could somehow gain control of additional mining-company stock. Damn, I'd best take Nolen's advice and not get greedy. Maida wiped her sweaty palms.

Glancing at her watch, she swore, realizing her repeated double-checking had made her late to return a scheduled telephone call. She wheeled her small car into the early evening traffic, driving until she spotted an unoccupied telephone booth. Maida wished to use her car phone, but she knew it was far too dangerous. She turned up her coat collar, as she stepped into the cold, gusting wind, blowing inland off the white-capped San Francisco bay. She hurried into the phone booth and pulled the door shut, unhappy that it barely sheltered her from the night chill, while she dialed Pam's number.

She encouraged herself to exude a happy tone of voice, even though she preferred to avoid talking with the reporter. "Pam, how are things in the glamorous world of bright lights and movie stars?"

"Fabulous, of course," the reporter replied. "Tell that handsome Nolen, that I spoke with book publishers in New York, and with several movie firms here in Hollywood. Thanks to my charm, I bewitched executives in each of the agencies. They expected an interview for a TV documentary. It never hurts to tell a teeny lie or two in this business. They thought I had lost my mind, when I offered to sell book or movie rights to a sensational story, about which I wouldn't supply any details."

"I'm sure you knew exactly how to handle them."

"Of course. Would you expect any less from a professional? I anticipated their severe skepticism. Nonetheless, I left a contract with each executive, after telling them an Association lawyer would be in L.A., and another in New York, when news of our story is released. Then, those same executives will jump at the chance to purchase our material. Bidding should go as high as two million dollars for movie rights."

"When I hear that type of news, I am thrilled about our plans," Maida replied.

Pam attacked. "I told Nolen you are too young, and too inexperienced, to be placed in such a vulnerable position. If you feel squeamish about this adventure, perhaps you should quit the Association, and run home to mommy and daddy."

Maida gritted her teeth, refraining from snapping back at the woman who was baiting her on the other end of the phone line. "I'm not quitting," Maida firmly answered.

"CMR would literally love to get their hands on you," Pam continued. "You are a key player in our efforts. You know what everyone in the Association is striving to accomplish, their location, and how much progress each has made."

"I'm not concerned about CMR. And Nolen is pleased with my performance."

"No doubt, Maida. I'm sure you don't hesitate to perform well for him—especially when you two are alone."

Maida changed the subject, though seething from Pam's persistent innuendoes. "Pam, remember to attend the Association meeting on the tenth of next month."

"Never fear, I'm looking forward to seeing Nolen again. Now child, you be careful. We wouldn't want anything to happen to you. Now, would we?"

Maida controlled her urge to slam down the telephone receiver. Instead, she stuffed her notebook into her purse, then jerked open the phone booth door. "Damn witch," Maida swore, venting her temper.

A cold shiver whispered up her spine, as she walked through the black windy night. Dull pain throbbed in her forehead, and her shoulder muscles ached. All she really wanted was a hot bubble bath, an undisturbed night's sleep, and freedom from worry over a myriad of details.

For a moment, after slipping into her car, Maida closed her eyes. She fantasized about being alone with Nolen, lounging on a vast, white beach in Mexico. How wonderful it would be, she thought, strolling hand in hand, happy with each other, no schedules to meet. We wouldn't have CMR stalking us. We wouldn't be concerned about the time needed to complete everything before the filing. Time, she sighed, the one relentless element our Association cannot control.

21

Traps

Tower stomped into his office. Small red splotches highlighted his face. His assistant trailed into the room, as the executive ripped a cigar from his desk. He bit off the end, and spit it toward a brass wastepaper basket, but missed. Tower jumped toward the receptacle and gave it a savage kick. The metal container flew several feet, crashing into an expensive cloisonné table lamp, knocking it to the floor. "Lousy shit!"

"Trouble at the morning meeting, Uncle?"

Tower answered by lunging at the vulnerable basket and kicking it again. "That SOB from the Anchorage office isn't going to steal the Pacific Division Presidency from me. He subtly needled the CEO till I got my butt chewed for the poor semi-annual exploration report. I believe they expect me to shit gold and piss oil!"

The assistant scooped the broken lamp chips into a file folder, and discarded them into the reconfigured wastepaper basket, while his uncle dropped into his leather chair. Tower hit a button on his phone console which automatically dialed Boodan Tribou.

"Boodan, it has been two and one-half weeks since I put you on my payroll. This isn't a welfare office. I'm paying you for results. If that piss-head Martin brings in a major gold discovery in our exploration region, without CMR having a piece of the action, then my ass is grass. What have you accomplished?"

"We're doing great. You ain't got nothin' to worry about, cousin. You'll get credit for the strike, and be top dog in the mining industry this year."

"Bullshit! Don't hand me inflated platitudes. I want facts —real action!"

"OK, I'll give you the latest information, face-to-face. I'll jet down and lay out how I've boxed-in Martin."

"Are you going deaf?"

Boodan extended the phone away from his ear, as Tower shouted.

"I said facts, Boodan. Now!"

"All right, we'll do it your way. I saw Mrs. Jorgenson this week. Told her that her boy had hit a valuable deposit. Told her he was going to keep the wealth she and her husband had worked all their lives to find. She wasn't buying my story, until I showed her a copy of the gold assay, with his signature. Asked her why hadn't he told her about the strike? He knows she's sick and still hurtin' over her husband's coma. After awhile, she came around; cried a lot."

"So what? Do you think I care about some old woman's tears?"

"T.J., if we have to, we can claim Digger Jorgenson pinpointed the site first, but was hospitalized before he could file. So, Mrs. Jorgenson is the rightful owner. I got her to sign a power of attorney, allowin' CMR to represent her. I'd say we have a legal basis to challenge any claim Martin submits."

"Not enough. That move will only buy us an expensive, involved legal battle."

"I also had a team visit Martin's wife."

"You mean ex-wife."

"No, it ain't final yet. There was supposed to be one more court appearance in Virginia. But Martin didn't show. And lucky for us, she wouldn't finalize the divorce until he appeared. She's real bitter over the break-up. When she learned there might be a gold discovery, and her possible rights to some of it, she got excited and downright helpful. Anyways, we got a contract with her, giving us a second shot at Martin's mine. For our legal help, CMR will receive half of all gold shares or profits she gains from the divorce, or through inheritance."

The phone in the Cajun's hand fell silent for a moment.

"Well, Boodan, you saved yourself. Damn fine work!"

"I said you didn't have nothin' to worry about," Boodan continued. "With the help of Digger's old lady, the wife, and that phone book we stole from Martin's house, I made a list of people who might be helpin' him. I've had ten people trailed or looked for. Eight are definitely not involved. We still have one active tail, and one person we can't find."

"Who is the missing person?"

"Martin's brother, Paul. He's been traveling since two days after the killing at the gas station. I don't believe it's by chance, that Paul Martin disappeared right after his brother left the Susanville area."

"And who are you following?"

"My men are tracking Maida Collins, a stock broker in Oakland. Martin's wife claims he had an affair with this gal. After one of my men found Martin's motorcycle, hidden under a tarp in her garage, we've been keeping close tabs on her."

"Hell, let's kidnap the broker. She'll know where Martin is."

"I doubt it. So far, Martin's showing smarts. He probably keeps movin' so no one can find him. I'm bettin' he ain't told anyone the location of the gold either. I wouldn't, if I was him. Anyways, it's better for us to watch her. If they are having an affair, some night he will return to screw her. Then, we can grab him. And we have other advantages, even though we're not sure where Martin is, or exactly when he'll make his move."

"What advantages are you babbling about?" Tower questioned.

"Sooner or later, he's gotta' show up at the County Recorder's office in Susanville. By law, he has to record his claim in the county where he made the discovery. 'Cause of that, my men bribed the County Recorder. She'll delay any one tryin' to file a gold claim, so we can get there before it's final. Also, she'll give us the recorded map coordinates of any filing. I have a team in Susanville waiting to ambush Martin. If he's seen, we'll nab him. I guarantee I can wring the site location out of him in twenty-four hours."

"It appears you are spending my money wisely."

"There is another advantage. Martin knows we're aware of the strike. He's gotta' realize we're searching for it. He can't wait too much longer, 'cause we might find the site. And now, we can close in on it. The police report showed Digger's crash happened on Route 395, due east of the town. Both Martin and Digger have only been seen traveling east of the city. Along with what that assayist said about trace salts, there's confirmin' data that the gold site must be on the Nevada side of Susanville. Can't yet tell you if the mine is north in the Skedaddle Mountains, or south in the Diamond Mountains. But that mine's east of the town, that's for damn sure."

"OK," Tower replied, "I'll re-direct four of our field crews to begin searching east of the city. Then, we will have all eight teams between the town and the border, working the foothills near Highway 395."

"I've got more," the Cajun stated. "Two days after you hired me, I paid a gal to go to the cops. Had her claim she was drivin' by the gas station on the sixteenth of August, when a truck pulled in front of her. She told them she remembered the license number and Martin's description. Said, he almost sideswiped her, as she was headed to a birthday party in Susanville. Then, I got your lawyers to start pressurin' the state police to arrest Martin for the murder of one of CMR's employees."

"Great, I would love to put Martin's ass in jail."

"So, boss, we've got several traps ready for Martin, including plenty of legal weapons to use against him. Plus, I have a few other trump cards up my sleeve. We can play them later, if we need to."

"Now that's the type of material I enjoy paying for," Tower replied. "Fly down today. Dictate a two-page report to my assistant. List the deal with the old lady, as well as your other accomplishments. Ensure you make them sound legal. I'll escort you to see the company President. Then, I will explain to him that we have been closing in on a big strike, but we did not disclose any information at the quarterly meeting for security reasons. That should eliminate the heat for awhile."

"Sure. I'll make that report sound real good."

Tower tossed the phone back into its cradle, thought for a moment, then spoke to his nephew. "Schedule a late afternoon appointment with the company CEO. Tell his secretary, it's urgent."

"Possibly, a response to the questions he raised during today's briefing, sir?"

"Hum...yes. Yes, that should be vague enough to avoid arousing inquiry from one of his corporate entourage." Tower smiled, contemplating the coming meeting. "I'll present my position, so the President recognizes just how close we actually are to this gold discovery. Of course, I'll subtly accept all the credit. Simultaneously, I will reinforce what a buffoon that Alaska Division VP is. OK, get things moving."

"Absolutely," the assistant agreed. He stepped from the room.

This calls for a celebration, Tower thought. He flipped his desk telephone index to a white card, showing only four penned letters and digits. They were his code to enjoyment.

I shouldn't chance this again, he worried. It's so dangerous. If some business associate sees me enter or leave the bordello, they might uncover my secret. He savored the exciting memory of his last visit to the three-story Victorian house, feeling a bulge expand in his trousers. Hell, who's going to see me? I'm always cautious. Anyway, what's the purpose of being successful, if I can't spend my money tending my vices? Without further hesitation, he dialed the number.

A sultry, young woman's voice answered. "Hello, how may I serve you?"

"Is this Madame Debra?" Tower whispered.

"Can you be more specific?" the seductive voice quizzed.

"This is member A7H5," Tower responded, as he rubbed his crotch, waiting for the woman to enter his code into her portable computer.

"It's most flattering to hear from you again, T.J." the tempting voice sang. "I did not expect you to return so soon. You must have enjoyed your last session. Would you care to indulge yourself in the same manner this time?"

"No, I want a..."

"T.J., please speak louder. I can't hear you."

"I want a slave."

"Male or female?" she nonchalantly inquired.

"Could I...could I have one of each?"

"You realize, we do charge extra for dual submissives?"

"They must grovel and not complain when I whip them, or I will refuse to pay."

"Certainly T.J., but you must control yourself. We can't have another incident like last time. Remember, you went too far?"

"I will not go too far," he asserted.

"When would you like to schedule your session?"

"I can arrive by seven tonight. They must do everything I order!"

"Excellent. Randy and Sandra will be waiting on their knees at the front door to receive you. Enjoy your erotic evening."

Cholo Cantera relaxed in his Sparks, Nevada motel room, his hands interlaced behind his neck, thinking to himself. He compared the facts he had gathered during the past five weeks, against the wiretap information which the monitoring team had just delivered. I can feel in my gut that I've collected sufficient pieces of this puzzle. Enough to develop a clear picture of what's going on. I only have to arrange the parts, while not letting my biases block me from seeing the pattern.

The young agent opened his briefcase. He began sorting his notes, transcripts, and documents into orderly rows across the bed. Next, he reviewed all the important facts.

What do I know? Martin allegedly stole arms and ammunition to supply some right-wing extremist group. That act worries the government, because he's had extensive military training and anti-terrorist field experience. Unrelated to the weapons larceny, Martin made a promise to his dad to make one last prospecting attempt. Subsequently, he left his home town four weeks ago, to begin his search. One week later, he's involved with a murder of a CMR man near Susanville. The next day, he withdraws $4,000 dollars from an Oakland bank. Then, he disappears, after storing his motorcycle with Maida Collins.

Additionally, a CMR Vice-President is so convinced Martin has discovered a gold mine, that he has a private team intent on stealing what Martin has found. They even burglarized his home. Which is where I found the computer printouts displaying the Susanville coordinates. Finally, everyone I spoke with, as well as every past record, indicates

Martin is an honest individual, with no connections to any political group.

Over and over, Cantera attempted to determine any possible motive Martin might have for stealing the government munitions. Like an irritating splinter, the lack of consistency between the theft charge and the collected facts, could not be ignored.

Cantera paced back and forth, frustrated. Instead of assuming Martin is guilty, he argued to himself, what if he were not involved with the theft? He snapped his fingers and smiled. The pattern is finally clear. I know what Martin and CMR are doing. I'm just missing the last third. Why is the government using falsified data to gain control of this guy? It doesn't seem reasonable that someone in the Bureau would frame an innocent man. Is it just a terrible mistake by the government? Or does my boss for some mysterious reason want to know what Martin is doing and where he is? That sounds crazy. I'm acting crazy!

But why did the SAC refuse to discuss the information gaps apparent in the intel report? Well, I'm going to determine who's guilty of what, while trying to get my hands on Nolen. First, I'll find him. Then, I'll figure out what's causing the government to be concerned about an unimportant prospector.

I'll begin by authorizing a tap on that stock broker's office and home phones. Next, I'll track down the source of the murder report in the Susanville area. After that, I'll contact the CID in San Francisco. They should be able to more fully explain the weapons theft. And this time, I won't ask permission to do the investigating. Meanwhile, I'll submit a partial report, to keep the SAC pacified.

Cantera collected his notes into a neat pile. He retrieved a California road map from his briefcase and unfolded it on the bed. Ruefully, he smiled when he spotted the military installation sprawling across the high-plateau desert, east of Susanville.

Hell, he worried, what if I am wrong? Perhaps, the gold strike is just a cover, to allow Martin and some radicals to recon the depot? Maybe some idiots are planning to seize more munitions? This is becoming more and more confusing, not less confusing.

4 September

The Depot Commander broke the pencil he held and whirled his chair to the right, scrutinizing a wall map, displaying the county where the depot was situated. "Major, are you telling me this asshole Martin probably discovered a rich gold strike within eight miles of our laboratories?"

"Yes sir," the security officer replied. "The FBI wiretap, supplied by the Sacramento SAC, confirms CMR believes he did. The mining company is already searching for the site."

Rubbing his aching shoulder, the commander contemplated the message. "Christ, if the news media gets a whiff of this information, we'll have people swarming around here like locusts. Overnight, one-tenth of San Francisco, maybe more, could migrate here. People hoping to get rich quick, most of the unemployed, and even young kids seeking adventure, would race here. Imagine all the motorcycles, four-wheel-drive trucks, campers, even dune buggies roaring all over the hills and desert, day and night. There's no way we could keep the hordes from poking around the depot. Even our own soldiers might run off. We couldn't finish production or complete dismantling the labs."

"That scenario does seem likely, sir."

"What's the status on our next shipment?"

"Sir, the unmarked 737 jet transport is scheduled to touch down at 2:56 a.m., tomorrow night. Its destination is the Royal Air Force base at Upper Hereford, England. After that flight, we will have only one more production run and one final shipment."

"Who else around here knows about this report?"

"Only you and I, sir. I have been decoding all messages transmitted between us and the SAC. That way, we minimize the number of people who know FBI personnel are involved in locating Martin. After reading this transmission, I came directly to you."

"Good work. Burn this hard copy, and keep this information TOP SECRET. Do not tell anyone else! Understood?"

"Yes sir."

The Colonel began nervously pacing. "I'd say we have three situations which may occur. First, Martin does not release any word about the strike, until after we fly out the last shipment, four weeks from now. Or, damn his soul, he does release the information before our next flight, forcing us to move the last load by ground means. Or, we stop Martin from releasing any notification, long enough for us to use the standard airplane procedures."

"Yes sir, those are reasonable possibilities. However, the last one requires locating Martin. And that has been an impossible task, so far."

The Colonel winced, as he experienced a painful spasm in his stomach. It was one of many he had suffered during the last several weeks. "OK, I'll order Operations to prepare the ground movement option, then meet you at the communications vault. I want to talk to our FBI counterpart. The SAC needs to get serious about arresting Martin for the murder of that man at the gas station."

22

Town

5 September

Nolen tried ignoring the persistent, far-off voice urging him to wake up. He shifted his body to find a more comfortable position for his head, to relieve the awkward strain on his neck. Half conscious, he felt the chill of the car window pressing against his cheek. He rolled from it, annoyed, seeking a warmer spot in the cramped bucket seat.

Again the voice harassed him, intruding on his rest. Nolen wished the noise away, longing to return to the pleasant euphoria of sleep, which he sorely needed. His mouth gaped open. Only when he was exhausted, did he sleep that way. Abruptly, a hand clasped his shoulder and shook him, disregarding his mumbled complaints, forcing him to open his eyes.

"All right, I'm awake," Nolen growled, as he yawned.

"That's what you said five minutes ago," Hamilton chided. "Come to your senses, and make yourself presentable."

"Yeah, yeah," Nolen grumbled, as he raised his head above the dashboard and squinted into the brilliant afternoon light.

The car was parked on the gravel shoulder of a straight, two-lane asphalt road. It rocked, buffeted by the cold, fall wind cutting across the flat valley. He viewed the nearby fields and small hills, noting they lacked any hint of green.

Not receiving rain in the last twenty-two days, had clearly left its mark. No animals stirred, not even a solitary bird flying above the dry, stark landscape. An emptiness, almost a loneliness, encircled the car. "Where are we?" he questioned.

"About five miles north of Susanville, a few minutes drive from the headquarters of the construction firm. It's on the back side of that low ridge, up ahead."

Nolen stretched, rubbed his bearded face, then checked his watch. "Well, I feel somewhat better than I did this morning in that real estate office, back in Nevada. Two hours of sleep, while you drove, almost made me feel human again."

The banker poured a steaming cup of coffee from his thermos jug, and handed the cup to Nolen. "The land purchases went exceptionally well today. Don't you agree?"

"Hamilton, the purchase went through quickly because of your prior friendship with those real estate agents, as well as the sellers. You're the reason we were able to buy two large parcels, adjacent to each other, in different states. You certainly reassured the sellers, by showing them the bank drafts."

"A noticeable attitude adjustment occurs when you place $60,000 into a person's hands." Hamilton beamed, as he styled his toupee.

Nolen sipped the hot, black liquid, then spoke. "I was pleased with how the California and the Nevada agents cut through all the red tape."

"Nolen, that came about because I had made an agreement with each broker earlier. I assured them, if they helped us rapidly and successfully conclude these purchases, we would use them later this year to sell all our lots."

"H'm, those incentives should motivate the agents, and supply you with a team to sell the land during the weeks when the Association is busiest. You can also use the Nevada agent to finalize the sale of the fifty-acre smelter site to our future mining company. Hamilton, what price should we hang on the town lots?"

"Between eight to ten times more that the $800,000 we're obligated to pay for the parcel."

"That will net a tidy fortune. Too bad we couldn't acquire that additional 300-acre farm, farther north. But I guess I shouldn't expect everything I plan to fall into place."

Hamilton started the car and drove down the winding road, until he reached the rust-stained mobile headquarters of the Hooper Construction Company. To the left of the trailer, a sandy parking area was scattered with idle bulldozers, graders, and dented dump trucks. To the right of the headquarters, a small white house contrasted with the hulking, yellow earth-moving equipment.

As the car rolled to a stop, the usual afternoon wind gusted, creating a man-sized dust devil. It danced across the hard parking lot and into the nearby brown field, crumpling to the ground when it lost its invisible power. The sudden appearance of the small storm reminded Nolen how finding the gold vein tossed his life into turmoil. Hopefully, the Association would not die as easily as the tiny tornado.

Inside the trailer, the two partners negotiated with the owner of the firm. They repeatedly encouraged him to sell his company. Hamilton spoke in his loud voice, "Hooper, I think you're being offered a fair and profitable arrangement. I've been your banker for years, and you know I have always given you sound advice. When I heard Mr. Martin wanted to purchase a construction firm, of course I considered you first. Truthfully, the Board of Directors at the bank has been pressuring me to foreclose on your firm."

"Foreclose? No one said anything to me, not since I refinanced my equipment loan."

"Hooper, you have a $350,000 note. You are behind in payments by $50,000, and your company has only one small active contract. Let's face it, you're in trouble. Do you want your family to sacrifice everything you've worked so hard to gain during these last ten years?"

Hooper slumped lower into his simulated leather chair, as he assessed his situation. He pulled his chin down to his chest, tightening his lips. "I sure don't want to lose my company," he said, while glaring at the two men confronting him.

"No need to," Nolen assured him. "We've written the contract keeping you as manager of the business, guaranteed for two years. I'll pay you $36,000 a year in salary, and put up $425,800 for seventy percent of the company. We'll be partners. During the next ninety days, I'll provide $500,000 worth of construction contracts. You'll receive a percentage of the profit gained from each job. Which means, you'll be free of debt, continuing the work you enjoy, and earning income to boot! What do you say?"

"I need my lawyer to review this deal."

"That's crap," Nolen snapped. "I don't have time to waste with some damn lawyer. Your banker is here; he has explained the contract to you. Your attorney can review it later. You make this decision now. Take it, or leave it. Yes or no?"

The man gritted his teeth, then reached to slide the curtain away from the window. For a long moment, he stared across the dirt yard into the kitchen of his house. He watched his wife feeding his infant son. With mixed emotions, he signed the agreement laying in front of him. "Done," he grunted.

Nolen took three sheets of paper and a diagram from his briefcase. He slid them in front of his new manager. "Hooper, within five days, begin building a town at Tiller Junction."

"A what?"

Nolen continued. "It must be completed by September 28. Understand, you'll have only eighteen, twenty-four hour construction days to conclude this project. The job is structured in three phases. Phase one will encompass about $150,000 worth of work. Construct railroad and truck off-loading ramps, while you bulldoze an airstrip. The train ramp must be finished first, because I'll be shipping you most of our supplies by rail. We'll use the ramp to rapidly unload the material from the boxcars."

"No train stops at Tiller Junction," Hooper argued, still struggling to believe the surprising stranger.

"Don't worry, Hooper, Southern Pacific is already committed to contract. One train will stop every other day. The airstrip is the second priority. My partners and I will often be flying in and out of the new town.

"You must also establish a road network, level parking areas, dig a refuse dump, drill two water wells, lay pipe lines, pour cement slab foundations for all buildings, and surround the town with a perimeter fence."

"Hold on," Hooper resisted. "That's going to require all my equipment, plus far more. I'll need to hire many more men than I've got on payroll."

"That's what the next five days are for. There are enough unemployed workers locally, that you can easily maintain day and night crews. Bear in mind, I won't tolerate any labor disputes. Sign all employees to a sixty-day work agreement. In this file, I've included a schedule of wage rates. You're authorized to pay bonuses, if the work meets standards and is completed ahead of schedule. Hamilton has the payroll fund in operation at his bank. Tomorrow, he will open another account to accommodate leasing any additional equipment you may need."

The engineer contemplated the magnitude of Nolen's plan. "Yeah, it might be possible to do this. But it's certainly a farfetched challenge."

"One thing is vitally important," Nolen stated. "The water wells must be producing within fourteen days. Get the well-digging team started tomorrow or the next day, to afford us as much time as possible to finish that task. Advise me about any difficulties you have with hitting a large vein. Because, if we can't produce sufficient water from the wells, we'll be forced to haul it in from one of the local upland lakes."

"No water, no town; that's for sure," Hooper mumbled.

"Phase two will cost about $728,000. String power lines. Emplace gas, propane, and water storage tanks. And erect 27 buildings. Place half on each side of the border. The structures will be 200 x 200 foot aluminum sheds."

"Why the hell are we constructing a small city, way out in the middle of the damn desert?" the builder grumbled, still unable to fathom what notion was motivating Nolen.

"That's not your concern. Just build what I pay you to build."

"Yeah, well, you'll need lots of permits to do all this."

"I've already secured them. They're included in this binder." Nolen handed the folder to the doubting man.

Hooper scanned the floor plans and sketch map provided. "At least you're building a town that's functional, not fancy. That's wise, given the tight schedule you've imposed."

Nolen slid another sheet of paper across the table. "In phase three, we finish the interior of each building. This list details the structures and major installations inside each one."

Quantity & Type of Building	Interior
1 Administration/headquarters	Office equipment
1 Worker's Quarters	Bunk beds, wall lockers, etc.
1 Casino	Gaming areas, tables, bar
3 Bordello Buildings (small)	Beds, toilets, closets, etc.
2 Stores	Shelves, counters
2 Warehouses	Storage racks, work tables
2 Bars	Bars, tables, shelves
2 Flop houses	Beds, toilets

4 Chow halls/Theaters	Tables, benches and kitchen
3 Bath houses	Large hot tubs, lockers
2 Pump houses	Pumping equipment
1 Theater	Stage & Bleachers
2 Banks	Vaults, counters, desks
1 Airplane hanger	Work benches, office equipment

"This is impossible!" Hooper flared. "I can't gather all this material in two weeks."

"Don't sweat it. My brother has several buyers purchasing the listed items. Delivery will come by rail or truck from San Francisco, Sacramento, and Reno. Yesterday, I purchased a trucking company, which will transport all our material. The firm has eight semi-trailers, four 10,000-gallon gasoline trucks, and two 5,000-gallon water trucks. The buyers and shipper will work for you, according to my master plan. They'll both call you later today, after we furnish them your telephone number. So, we should not incur any scheduling problems."

"Son of a gun," Hooper swore as he grudgingly let a smile emerge. "This reminds me of the Navy."

"The Navy?" Nolen questioned.

"Yeah, I was in an engineer battalion. We did these rapid constructs all the time. Move in fast, bulldoze the countryside, and tilt-up pre-fab buildings overnight. Of course, then, I wasn't in charge of the whole operation. Where's that time schedule?" He shuffled the papers, placing several pages side by side. "Possibly, we can do it. But some adjustments are necessary. What about toilets for my crew?"

Nolen replied, "Two hundred mobile johns will be brought in for use during construction. Later, they will support the town, once it's operational."

"For a place this big, that's a good idea."

"Hooper, in two days, furnish your detailed material needs to the buyers. I doubt there will be many differences from the list I've provided."

"Damn, you're gonna' need a fat wallet for this one."

"Each building runs around $45,000. The payment terms are twenty-five percent down, another twenty-five percent twenty days after material delivery, and the remaining fifty percent is due thirty days later."

Hooper looked up from the plan spread on his desk. "Well, all this could be some real fun. And it's definitely going to be profitable for me. But why are you hell-bent on building a town? No one lives out there. No one will stop at the casino. And no one will frequent the bordello. Anyway, not enough to make it worth the money you're spending. You gotta' be nuts!"

"I may be, but remember, I'm paying. Because I am, there are a few critical ground rules you must follow. I want it done my way and completed on schedule. My brother will provide you with day-to-day instructions. Periodically, I'll visit the town to inspect the work. I've also selected the accounting firm which will handle payroll and material payments. Be aware, a separate firm audits the books monthly, as well as monitors the cash flow."

"Worried I might steal something?"

"For the next three to four weeks, you keep the lid on this project as tightly as you can. Neither your crew nor their wives need babble about what is being built. Tailor the labor contracts so the workers keep their mouths shut and sleep on site. Do we understand each other?"

"What would you have done if I hadn't agreed to sell?"

"Hamilton had identified an alternate company, also meeting our needs. We would have bought them out instead. Meanwhile, the bank would have called your loan."

The engineer scrutinized Nolen's steady gaze, and considered the stranger's determination. "Mister, this appears to be the beginning of a profitable partnership, at least for me. But our work is cut out for us, on this project. The only way your plan will succeed is if there are no construction glitches, failed deliveries, bad weather, or labor shortages. You nervous yet?"

23

Difficulties

10 September

Five days later, in a motel 120 miles west of Susanville, Nolen stood behind the wet bar joking with his partners, as he poured champagne for them. Excitement and laughter filled their conversation, as they lounged on the leather sofa, occupied stuffed chairs, and shared their adventures.

Nolen analyzed each associate, considering how the exhaustive work and extended hours over the last few weeks had changed each of them. Paul, for the first time in years, Nolen thought, has shed some weight. Well, that won't hurt him. Hamilton also looks healthier, with his newly-acquired tan, a bonus from working outside his bank. He still speaks loud, maybe too loudly. I wonder if something is bothering him? Pam hasn't changed, still hiding behind her camera and her perfect makeup. When she thinks no one is watching, she examines people like a hungry hawk spying a field mouse. Maida's allure is still prominent. But, she is more fidgety, and seems quieter than her usual talkative self. She might simply be tired. Her eyes have lost some of their unique sparkle.

Suddenly, Pam's camera flashed, as she snapped an unexpected photo of the group. Nolen glanced at a mirror, checking his own features. Fella, you better not appear tired or nervous around your partners. You must keep them calm and confident!

Nolen called out, "Good news, gang. I will file the claim in eighteen days." The small group cheered and clapped. Nolen continued, "I want everyone to explain what they have accomplished, so we all comprehend the big picture, and avoid any approaching pitfalls."

Nolen signaled to Pam. "What's the status of your efforts?"

The newswoman set aside her automatic camera, then rose to face the group. "I have contacted three movie and two book publishing agencies, to offer them the opportunity to bid for the rights to our story. And I've written the first three news releases, magazine stories, and electronic scripts. They are ready for Nolen to edit and approve."

She smiled. "The pieces are dynamic, emotional, insightful. I should win a Pulitzer prize because they are so good. Be assured, I will place us in the limelight, beginning the day we file the claim. Simultaneously, an Association lawyer will be positioned in Los Angeles and another in New York. They will consummate the movie and book contracts when news of the gold rush hits the major cities."

Hamilton stroked his toupee and asked, "Don't you agree I'll look dashing on the screen?"

Pam sat next to him and poked his belly with her elbow. "You're too fat." Everyone laughed when the banker pretended to be insulted.

"Now, Handsome," Pam said to Nolen, "tell us what you have accomplished."

"My time's been consumed orchestrating this whole affair, while establishing the legal structure of the Association. Coordinating the birth of our corporations, licenses, and permit agreements has kept me busy. We now have thirteen corporations. Three are headquartered outside the States, and ten are U.S. firms."

"Give us some details," Pam encouraged.

"In the Cayman Islands, we opened a fake holding company, together with the real holding company, which controls ownership of each Association business. The actual holding company will shield our profits from IRS scrutiny. In Panama, another overseas company has been set up to purchase and sell all Association land parcels.

"Five of our businesses operate within the new town. They are the Association, bordello, casino, investment, and real estate companies. The remaining firms are located outside the town but support it. They are our accounting, legal, construction, trucking, and marketing entities."

Nolen placed his hand on Maida's shoulder. "Please, continue with your update."

"Here in Red Bluff, over in Susanville, and in Reno, I've opened and staffed five accounting firms. In the Bay area, I've been purchasing stock in the mining company that we will retain to extract the gold. I've also acquired futures contracts, obligating us to deliver 10,000 ounces of gold within three months, at twenty dollars less than present market price. Once we announce that a massive strike exists, the price of gold will drop substantially below what people agreed to pay. The speculators will scramble to cut their losses. They will buy their way out of those contracts to avoid paying the higher prices. That's when we will profit."

The group clapped and cheered.

Nolen pointed to his brother. "Tell us about the town."

"Boomtown, you mean." Paul pinned a map of the town's lots and streets onto the wall. "My primary objective is having a functional town erected by the time Nolen files. We've been building twenty-four hours a day, for two days.

"When I first viewed Nolen's plan, I realized we needed the means to quickly separate the prospectors from their money. The majority of my time has been spent designing that production line. A line, where people flow through the town as rapidly as we can move them, leaving their dollars behind. So far, I've signed thirty-five contracts for food, dry goods, prospecting tools, gasoline, and building materials. Construction and other supplies are now funneling into Tiller Junction, both by rail and by truck. I have been doing some wheeling and dealing!"

Paul pointed to individual buildings outlined on his map. "The town provides the necessities that gold hunters need and want. I mean food, water, lodging, liquor, mining supplies, gasoline, gambling, and sex. Following the gold strike announcement, I estimate 80,000 men and women will flood the region within the first forty-eight hours. Because we are near the strike, we're guaranteed to be quickly functioning at maximum capacity. My calculations indicate that the town can handle 10,000 people a day. That's when railroad deliveries will become invaluable, because the surrounding roads will be clogged with bumper-to-bumper traffic."

"Ten thousand people will generate an abundance of money each day," Pam remarked.

"My intent is to keep the prices low for meals, baths, and sleeping facilities. Such inducements will draw people into the town, fund our overhead, and generate higher sales on everything else we offer. I estimate the average prospector will remain in town at least one day. He'll eat two meals, enjoy a bath, and buy a basic set of prospecting gear. He will also consume three drinks, gamble for at least thirty minutes, and get laid once. All, for a total of $246. That alone will generate $2,460,000 each day that we are in operation. Should the gold rush last ten days, we'll take in about twenty-four million dollars. Fifty percent of that income will be sheer profit. If we raise prices higher, we'll make even more. I plan to push the prices as fast and as high as demand will allow."

Maida asked, "How many workers will we employ, in the town?"

"We'll run a 750-person work force—two twelve hour shifts of 375. One hundred and five of those will comprise the security force. Hopefully, they will be effective in maintaining order, as well as guarding our buildings."

Paul paused, as he pointed at Nolen. "Two days before filing the claim, 210 prostitutes will be bussed into town. It hasn't been an easy task, coming up with so many women for the bordello. At least, not without letting the world in on our little secret. I've kept one man busy full-time, soliciting women from Seattle to San Diego, out to Phoenix, then north to Salt Lake City. We hired no more than twenty whores per town, and stayed clear of Nevada, to avoid letting organized crime learn we are establishing a large prostitution operation."

"Do we need to worry about the Mafia?" Maida asked, somewhat unnerved.

Paul wiped a shimmer of sweat from his forehead. "Well, I'm trying to avoid that problem, until after we are operational. For added protection, we tell each prostitute the same cover story. They are being hired for two weeks, to service rich West Coast clients, visiting a dude ranch. Most women will choose to stay longer than fourteen days, particularly when they realize how much money they can make by remaining with us. Hell, some day I may be paying those gals a pension—if the mob doesn't rub me out!"

"Who will manage these women?" Pam questioned. "None of us have that kind of experience. We might lose hundreds of thousands of dollars to these greedy women."

"That's why I hired a Madam from a cathouse in San Francisco, to operate the business. She has all the red-light savvy we could ever wish for."

"Must we sell sex to the gold rushers?" Maida asked. "I had hoped the bordello plan would fizzle. I'm not comfortable with the prospect of exploiting women."

"Look, Sweetheart," Pam responded, "these women will substantially contribute to the magnetism of the town. If you had thought it out, you would realize that other boomtowns will spring up overnight to compete with us. They will also offer sex for sale. We, at least, will afford our female staff safe, decent working conditions."

"Safe working conditions, perhaps," Maida snapped. "Decent is not the operative word, here."

"Honey, if your principals are so strong," Pam snipped, "I'll gladly accept your share of the bordello profits."

"Ladies," Nolen interjected, "let Paul detail the status of the other ventures in the town."

Paul quickly continued. "Our two cafeteria buildings will provide meals four times a day: breakfast, lunch, dinner, and midnight. No menu choices. A patron receives two pre-prepared trays, containing a hot meal and plastic utensils. That is the quickest method to move people through the food lines."

"Sounds just like an airline meal service," Maida added.

"True, except we supply much larger portions on each tray. Enough starch and carbohydrates to satisfy hungry men.

"For the baths, I developed one of my better ideas. We've drilled three wells. One is producing a light flow. Without sufficient water, our town will dry up. So, to lessen the amount we are forced to haul in, I devised a method of using large, wooden tubs—similar to Japanese baths. We can accommodate more patrons, use less water, and recycle it more easily, using tubs rather than showers."

Pam pointed to the wall chart. "What do these symbols represent?"

"We charge for everything," Paul answered, "even entering the town, at those gates. We also rent sites for RV parking. The warehouses, stores, and gas pumps are being built side-by-side, assembly line style, allowing vehicles to rapidly pass by to load mining supplies.

"I also petitioned the Gaming Commission in Carson City to upgrade the Tiller Junction gambling license. I paid $8,000 apiece, to two Commission members, in exchange for expanded casino operations. The only requirement levied upon us is a cursory background check on yours truly. The Tiller woman did not need one, since she had a thorough investigation when she took over following her dad's death."

"Are the gaming people becoming suspicious?" Maida asked.

"They just think I'm crazy. To them, Tiller Junction is bullseye in the middle of nowhere. They think it will never be a profitable site, nor will it ever provide competition for anyone elsewhere in the state. So, they believe it's harmless to modify the license, since its use is restricted to the county where the bar is situated. I'd love to see the expression on their faces when they learn a gold strike was discovered within miles of our casino!"

"Could their surprise cause us problems later?" Maida questioned. "Will the Commission revoke our license? Or might mobsters, from one of the existing casinos, force their way into our gambling business?"

"I've been working hard to avoid such challenges. We're scheduled to receive the new license within a week. Our contract lawyer will witness the signing, to ensure the license is legal. We must obtain it before the gold strike announcement, or the Commission will blackmail us for millions."

Paul wiped his sweaty hands with his handkerchief, as he continued elaborating. "The casino will be bare bones. Our business lawyer is purchasing gambling equipment from several different sites around the state. His goal is to avoid arousing anyone's interest. We'll own fifty slot machines, ten blackjack tables, ten crap tables, and five roulette tables. The state inspector will check our casino operations during the week of the filing. I'm certain that will cost us another bribe."

"Paul, you seem tired," Maida commented.

"The pace has been brutal," Paul agreed. "If it continues, I will accomplish six month's work in six weeks. I do need some sleep."

"If this town flourishes into the money-production line you envision," Maida questioned, "how will we protect all the cash we accumulate each day?"

Paul nodded at Hamilton across the room. "Explain the bank procedures."

"We will use two temporary banks to secure the earnings each day. One bank on each side of the state border. Security guards, carrying shotguns, will be stationed inside and outside each temporary bank. If a gang tries to rob us, they will probably hit only one of the banks. So, we would only lose half of our daily income. Each evening, our money will be flown to one of ten different cities within California, to be placed into a more secure bank vault. It's a neat package, with quality protection for our money."

Nolen again spoke. "Hamilton, also reveal what's transpiring with the land sales."

"Well, we've completed all our real estate contracts, together with all down payments," he answered. "The Association owns adjacent property along both sides of the border. We control a total of 550 acres surrounding the town. Plus, another fifty acre piece, farther north, suitable for a smelter site."

He walked to the wall, where he pinned an overlay on top of the town map. "The parceling design has been completed. When the gold rush begins, we'll sell one-half of the acreage in small lots, to fuel boomtown growth. By selling lots, our effective land re-sale price will be ten times higher than our initial purchase price."

"What is the situation with the other bank directors?" Maida probed.

"Everything is fine," Hamilton loudly assured the group. "A couple of the directors were a bit nervous, when they learned I had signed an agreement issuing millions of dollars of credit to the Association. However, I informed them only after we had established the corporation accounts. By then, it appeared the Association business arrangements would bring in sufficient money to outweigh the risks facing the bank. Maida had begun depositing checks from her clients buying mining stock. Also, Paul had started making purchases from businesses owned by the other bank directors. They were pleased they were benefitting from the Association's local activity. Everything is fine."

"Is there danger of a state or federal audit?" Maida inquired.

Hamilton adjusted his toupee and bit his upper lip before replying. "Well, yes, that could make our situation considerably more difficult. However, we have not violated any banking procedures. Nor have we done anything illegal. Besides, we just completed the annual audit three months ago. Another formal review should not occur soon."

Maida persisted. "If one of the other bank directors thought things weren't proper, couldn't he request an audit by the state?"

"Yes, that's possible. However, I am monitoring the other directors. If someone begins acting skittish, I may be forced to tell him about the gold."

Annoyed, Pam quizzed Maida. "Well, what is the cash status of the Association?"

Maida selected several papers from her attaché case, then pushed a stray curl of hair behind her ear. "Our original cash inflow was $149,000. It was deposited, together with the $1,500,000 worth of gold, allowing us to acquire a credit of $3,900,000. Consequently, our beginning bank position was slightly above five and a half million dollars—composed of cash, gold, and credit.

"We spent $79,000 for initial travel and start-up costs. Since then, we've had a continual outflow to buy land, licenses, material, and payroll draws. We attempted to structure all material and supply contracts on a ten percent deposit basis. That did not occur. Because we demand immediate delivery, we are averaging closer to twenty percent down. Thus far, we have spent a total of $4,900,000. This means, we have exhausted all our credit. Now, we are well into the principal." She frowned. "An audit team, or a concerned bank director, would find our spending versus replacement rate very unfavorable."

Hamilton nervously adjusted his tie.

Maida went on. "Plus, we are committed to contract payments of $3,000,000, due twenty-three days from now. And, another $6,600,000 due in forty-three days. Because our major income flow does not begin until we file in eighteen days, we have only five days slack, to receive enough cash to meet the twenty-three day commitments. That money should include the $200,000 advance on the sale of the smelter site, $2,000,000 worth of book and movie rights sales, combined with four-day's profit from the town, adding another $9,600,000. As long as we remain on schedule, our position is tenable. But if something delays us, we won't have to worry about the bank directors. We'll enjoy eighty angry creditors, carloads of police, and several state audit teams, swarming through our empty town."

Hamilton urged Nolen, "What about mining additional gold from the vein? From what you've told us, Nolen, you could extract another load in just a few days."

A scowl wrinkled Nolen's face. "That's a dangerous plan. By now, CMR has plenty of people around Susanville, searching for me and for the mine site. Because of the danger of losing the mine to CMR, or being prevented from filing, I don't want to be followed by them."

"Let me see if I understand the full scenario," Maida said. "We don't have enough water to fully operate the town. The Mafia may become very interested in our activities. We don't have our casino license finalized, nor gaming tables approved. And we are rapidly depleting our principal in Hamilton's bank, which may cause a disgruntled director to initiate an audit of our accounts. Meanwhile CMR is hunting for you and the mine."

"So what's the really bad news?" Nolen joked.

The group did not respond to his attempt at humor. Instead, all remained uncomfortably silent, considering the tenuous situation.

Nolen's green eyes searched the faces of his partners. Their jovial mood had evaporated, as the risks confronting them surfaced. Concern marked each individual, as he or she considered the strong possibility that their plan could fail, now that they were three-quarters of the way toward achieving their goal.

"Damn, we're in deep trouble," Pam blurted. Realizing she had stated aloud what everyone must be thinking, Pam hid behind her camera by taking additional snapshots. She photographed shining sweat beading over Paul's tired face and Maida reshuffling her papers. Pam ignored Hamilton's fingers drumming a nervous staccato on the tabletop.

Nolen interrupted their worries. "Yes, the easy part is over, and we are committed. Yes, you are beginning to feel the stress of betting your fortunes and your future on this gamble. But don't allow the pressure of the deadlines to unnerve you. If we bind together, we can succeed. Continue guarding the secret purpose for our town, as it takes shape and people start questioning why it's out in the desert. Don't worry. I have a way to protect us."

24

Theft

10 September

A half hour after the meeting, Nolen was alone in his motel bathroom when he heard a knock at the door. The sudden noise startled him. I'm becoming paranoid, he thought. Dwelling on the thousand things that can go wrong is making me jumpy. "Who is it?" he yelled.

"It's me, Pam."

"Hold on a second. I'll be right there." He wiped the remaining trails of shower water from his legs, then slipped on a pair of sweat pants. He verified, looking through the drapes, that it was the reporter and that she was alone, then he opened the door. "I wasn't expecting you. Is there some problem?"

"I have some information you need to hear." As Pam stepped into the room, she let her hand purposely slide across Nolen's bare chest. "Information that would have really soured spirits during our meeting. So, I saved it till we could be alone."

Nolen grabbed a pair of jeans and a sweater. "You can explain while I get dressed in the bathroom."

"No need to be modest. It doesn't bother me to see you half clothed."

"Thanks, but I was changing when you knocked. What's your news?"

"Are you aware, you are wanted for murdering the man in the gas station?"

"That doesn't surprise me. I assumed the police would rule out suicide, when they found him with his gut ripped open."

"What about the FBI claiming you're a member of a terrorist group which stole ammunition from the Army?"

Nolen dropped the sweater he was holding. "What?"

"An old friend of mine is a desk sergeant in Oakland. He let me review the All Points Bulletin issued for your handsome body. That's when I learned about the FBI involvement, as well as the terrorist issue. So, I went to Sacramento, where I spoke with the FBI Regional Director. He's real right-wing conservative."

"Did he think you were just another liberal reporter?" Nolen asked.

"Actually, his primary concern seemed to be my voluptuous chest. I purposely left the top three buttons of my blouse undone. He became fascinated with my breasts."

"Sounds like a unique interviewing technique."

"Men certainly continue providing additional information at those times. Anyway, this FBI character was extremely pleased that a reporter was interested in writing a story about you. He wants everyone along the west coast beating the bushes for your hide. He said you are a traitor, ranks you together with Benedict Arnold. It was wise of you to grow a beard. Perhaps you should also dye your hair. I could help you do that tonight."

"This terrorist story is crap! The government's using it as an excuse to locate me. They're concerned that I may have learned something about their secret operations. I'll just continue avoiding the police, and now the FBI, until we announce the gold strike. Then, I'll be able to prove what I was doing in the ravine."

"One good thing came from this turn of events," Pam remarked. "I led my boss to believe that a terrorist group is planning to sabotage the San Francisco water supply, with plutonium. And, because of your military experience, you were able to infiltrate their group. Furthermore, within three weeks, you will expose the plan during a Channel 25 telecast. Consequently, the APB helped me convince my boss that we have an important story, requiring me to be so secretive when contacting you. But I can't maintain this charade with my boss much longer."

"Maybe it would help if you unbutton your blouse when you're around him."

"Unfortunately, it won't. He's gay."

"Then think of something else as creative as the infiltration story."

"Don't worry. I can handle my end of this arrangement. But, Maida's comments about running out of money frighten me. We could all go to jail. Especially, if some greedy politician or local bigwig stumbles upon our scheme. They'd find real violations of the law by the Association, or invent false ones to ensure our arrest. They would file charges against us as a basis to void our claim. I'm telling you Nolen, I'm very concerned. Your plan is like a Chinese juggling act, and I don't want it tumbling down, squashing me on the bottom."

"Yes, costs are mounting higher than I expected. That changes things somewhat."

"What are you going to do? I agree with what you said during the meeting. CMR would love to trail you back to our vein."

"Well, they won't get the chance. Every good soldier maintains a reserve and knows when to use it. It's about time for me to do just that." Nolen scowled. "Look, I don't want to think about it anymore. I need to get some sleep."

"What you need is a pleasurable night's rest." Pam moved close to Nolen, so close that he could appreciate the subtle fragrance of her cologne. "I could be very cooperative, if you..."

Nolen interrupted. "I appreciate the offer coming from such a beautiful woman However, accepting would introduce complications into the Association that we simply don't need."

"Perhaps, you should reconsider." Pam wrapped her arms around Nolen and kissed him.

"You-hoo, Nolen, it's me," Maida called, as she unlocked the motel door, and pushed it open. "Oh, oh," she stuttered. "I didn't realize I might interrupt something so...so personal. Do you grope every man you know, Pam?"

Pam smiled as she reluctantly released Nolen. "Maida, we were not groping. The word is caressing."

"No doubt. You've probably caressed half the men in California."

"Only the best ones, Honey."

"Go to Hell!" Maida shouted, as she slammed the door.

"Damn it," Nolen snapped. "This is just what I meant by complications."

"Oh really? What is she doing with a key to your motel room?"

"That's none of your business."

"Nolen, that young thing can't begin to satisfy you the way I can. Let her have a cry. It'll do her some good. Anyway, you need me much more than you need her. Without me, your plan goes nowhere. Remember, I'm the one who's going to trigger the gold rush. Or perhaps I should release the story right now," she threatened.

"Pam, I don't need this crap from either of you two women."

"And I don't take rejection well! How the hell do you suppose Paul and Hamilton will react when I tell them the FBI is after us?"

"Pam, go back to your room before you do something stupid, just because you're angry. We don't need fights between partners endangering our efforts. I'm going to tell Maida the same thing. Call me in a couple of days, when you calm down."

Nolen left his motel room. Maida had crossed the courtyard and was entering her suite. He raced after her.

"Go away!" she shouted through her door.

"Maida, let me explain. We need to talk."

She opened the door, but did not unhook the chain. "What do you want? Didn't you get enough from that redheaded bimbo?"

"You're jumping to conclusions, Maida. You're the only one I've allowed close to me in months."

"What conclusion should I reach when I see you half naked, with that woman rubbing against you, giving you a tonsillectomy with her tongue?"

"Pam initiated the whole thing, not me. She had just approached me, when you came through the door. It's unreasonable for me to give you a key to my room, if I wanted to have sex with Pam. Sleeping with two women would ruin everything."

"You might have done it, if you are just using me. Maybe you are using all of us, pushing everyone's buttons just to get what you need. Or possibly you're like most men. After enjoying your conquest, you move on to a new thrill. Do you want to end our relationship?" Maida's voice cracked. She looked at the floor, so Nolen would not notice her tears.

"Maida, you're jealous, and you're worried that I don't need you anymore. Think, before this gets out of hand. I want our relationship to withstand interference from others. This will not be the last time people thrust themselves into our lives. This incident is just a taste of what will come. After we bring in the mine, men and women will clamor all over both of us. Now, and in the future, we will have to trust each other."

Maida hesitated, then closed the door. She unhooked the chain. "I want to believe you." She slipped into Nolen's arms, as soon as he stepped through the doorway.

"I apologize for becoming so angry," she whispered, as she squeezed tight against him, rubbing her tears dry on his sweater. "It just shocked me, seeing Pam kissing you. She may be a partner, but she's not my friend."

"It's OK, I understand." The couple sat on the end of her bed. "How have you been, pretty lady?" Nolen softly inquired.

"Lonely and tense. Some of my efforts go against my small town values."

"You're not going to hurt anyone. In three weeks, those people you make rich will be thanking you."

"That logic doesn't console my conscience. Nor does it chase away my fear of being caught by someone at my office, or by the SEC."

"Do you want to quit the Association?"

"I've often thought about leaving. I suppose I'm just anxious. I keep all my feelings bottled inside. I can't discuss what I'm experiencing with my family or anyone at work. Each Association member has been preoccupied with their own tasks, only talking on the phone or meeting at short gatherings, like today. We're not a family. We are a group of strangers, still questioning whether we can trust each other."

Nolen reached for Maida's hand. "You and I are not strangers. Why don't you show me how well you perform tonsillectomies?"

Maida finally smiled. She rolled him onto his back, then lifted his sweater. Her fingernails etched his muscular stomach. His body quivered with pleasure. "Please," she whispered, "don't embarrass me by making loud noises. These motel walls are paper thin."

11 September

Gray fingers of early morning light outlined the drapes in Pam's motel room. Gradually, the faint light pushed the darkness away, while the angry reporter lay under the rumpled covers of her bed. She stared at the ceiling. What did Nolen mean last night when he spoke about protecting the Association? Did he tell the truth or simply lie to keep all the partners calm? Well, I'm not calm. I'm worried about his grandiose scheme failing! There's no way he could have enough cash to cover the bills that Maida described. That little bitch!

Unless...unless, there's a rich silent partner in the Association? I doubt it. Nolen's been too careful, too fearful of losing control of his destiny. Damn him! This just isn't fair. While he's hogging fifty-one percent of everything, we minor partners do his bidding. And for what? Just for a measly eight percent of a dream that could fall apart for the slightest provocation! Either he's lying, or he's withholding something vital. Reserve? Could he have more gold hidden away?

Pam recalled the cramped scene inside the storage room, weeks earlier. She had been surprised and thrilled upon seeing the vast wealth hidden within such an ordinary trunk. There was the pickup, the scattered equipment, and the distracting foul smell drifting from the pile of stinking garbage. Garbage that smothered another trunk, one which looked much like the one Nolen had opened for her! Goosebumps scampered along Pam's arms. Her eyes widened. Perhaps the second trunk is the reserve he mentioned? Maybe it holds an additional $1,500,000 worth of gold? There could be an easy solution to my situation!

Pam jumped from bed and dressed. Hurriedly she packed, then grabbed her car keys, and rushed to her red hatchback sports car. She restrained herself, while driving through the small town where Nolen had assembled the Association. But when she wheeled onto U.S. Highway 5, she swept through the gears, and jammed the gas pedal against the floor. She raced south toward Oakland.

Three hours later, in Oakland, she screeched her automobile to a halt beside a twenty-four hour photo lab. She had often used the processing lab to develop pictures she had taken while covering a story. From her camera bag, she extracted the four rolls of film she had previously taken of the Association members. At the service counter, she tipped the clerk twenty dollars for immediate development.

Returning to her auto, she drove to a nearby gas station where she bought a city map, then walked to a phone booth at the corner of the lot. She thumbed through the yellow pages, until she found the section listing self-storage companies. The section filled four pages. She ripped the sheets from the phone book and slipped them into her purse.

Pam next searched the telephone book to find an equipment rental firm. A quick call verified a nearby firm rented bolt cutters. Twenty minutes later, tool in hand, she returned to the photo shop.

On the office counter, she unfolded the map and laid the telephone pages beside it. Using a green marker, she colored in two spots. First, the Oakland parking lot, where Nolen had stopped his motorcycle and blindfolded her. Second, the street corner where he later dropped her to catch a taxi. With a red ink pen, she began highlighting the location of each self-storage site in the downtown area located near the green marks.

She stepped back to look at the map. Thirty-five red spots measle-dotted the sheet, surrounding the two green marks. Damn, she thought, this looks impossible. I don't have time to visit all these sites. Somehow, I must speed this search. Where are the most likely places on this map to begin looking for Nolen's reserve?

She recalled traveling to and from the storage room, riding the motorcycle blindfolded, hanging onto Nolen. She remembered the frightening feeling each time they bounced over a bump in the pavement. She visualized Nolen, standing near his motorcycle in the store parking lot, explaining that the ride would only take about ten minutes. It seemed longer—more like a fifteen minute ride. And we averaged about twenty-five miles per hour. But I wasn't very calm during those two blindfold rides. My estimates could be misleading. So, I'll assume the storage bay must be within a ten-mile diameter circle, centered on the parking lot.

On the office counter she helped herself to a scissors and to a roll of string. She snipped a foot of cord, then used the legend at the base of the map to measure a five mile distance along the string. Placing one end of the cord at the parking lot mark, and holding the green pen at the other end, she scribed a circle on the map. She repeated the process, using the street corner as the center of the second circle. Twenty-one red spots were enclosed by the overlapping green rings.

Pam peered at the map. She reached into her purse and removed her cigarette case. It was empty. Frustrated, she crumpled the paper pack, and threw it to the floor. "Isn't that film ready yet?" she shouted.

The clerk stuck his head through the curtain, blocking the view into his processing area. "Miss Pendleton, you've utilized our services before, and should know that I work as fast as I can." He popped back behind the drop cloth before Pam could voice her impatience. Minutes later, he reappeared, carrying eighty three-by-five-inch, glossy enlargements.

Pam threw $200 onto the counter, before snatching the prints from the young man's hands. She rifled through the photos, retaining only those pictures of Nolen. Then she grabbed the map and telephone pages and hurried toward the doorway.

"Don't you want these other pictures?" the clerk asked.

"I'll get them later."

Pam drove to the nearest self-storage firm within the green circles on her map. There, no one recalled renting space to anyone who fit Nolen's description. Pam showed the pictures to employees at two other storage facilities, but received only negative responses.

I must use a better approach, she thought, as she sat in her car. "Think, girl, think!" She tried remembering every detail of that day when she first saw the trunk of gold. She began with the time they left the parking lot, and retraced the events up to the moment she stood looking at Nolen's shadowy figure. A shadowy figure, because he had stood in the painfully bright, open doorway. She snapped her fingers. "Blue and orange!" she exclaimed. "The exterior wall across the driveway was painted blue and orange."

She snatched the telephone pages from her purse. Yes, yes, here it is, she thought. 'Blue and Orange Storage, with six convenient locations throughout the East-bay.' Pam noted three of the branch firms were within the green circles on the map, and one just outside. She immediately used a sidewalk telephone to quiz the desk clerks at the storage sites. The third contact was rewarding. She learned they did have large bays; large enough to house a pickup inside, but that wasn't allowed.

Pam drove to the storage firm. Spying the blue and orange striped decoration painted across the walls, she became excited. She hurried into the manager's office.

A bulbous, heavy-set woman, wearing a bold black and white dress, stood and shuffled to the counter. "What can I do for you?" she asked with a strong east coast accent.

Pam opened her wallet to display her press credentials. "I'm Pam Pendleton, with Channel 25 TV news. I'm assembling a story about dead beat dads, and I could really use your help. Have you ever seen this man?" She held out two photographs of Nolen. One, she had taken shortly after meeting him, without his beard, and another from the night before.

"Well, Honey, I haven't missed a day of work in four years. So, if he came in, I would've seen him." The fat woman put her eyeglasses on and squinted at the photos. "He doesn't look familiar. Then, he doesn't look unfamiliar, either. Kind of a looker though."

Pam frowned. "He would have been here almost a month ago, around August seventeenth. He probably rented a large storage space."

"I'd sure like to help you," the clerk said as she turned to a file cabinet to remove a ledger. "Most men are shits. Least ways, all the ones I ever met. And any man who deserts his wife and kids is about as low as you can get." She thumbed through the book, running her finger down each page, then turned and stared at Nolen's picture again. She closed her eyes, hummed part of a tune, then smiled. "Yup, he was here! I remember now. Unshaven guy. Not real talkative, and sure in a rush. He rented Bay 121."

Pam reached across the counter and squeezed the large lady's hand. "You've been a real help. I'll mail you a copy of my article when it's published." As Pam left the office, she noted the windows had venetian blinds, adjusted to prevent sunlight from flooding the office. They partially blocked the view of the driveway. Pam left the storage compound and parked one block away.

After fifteen minutes, she returned, hoping the clerk would not be walking about the enclosure, or notice Pam's car returning to the area. Pam parked next to Bay 121. She watched a couple unloading furniture two stalls down from her. "Damn!" she swore, then lit a cigarette. She impatiently tapped her finger on the steering wheel, then stepped from the car and glared at the couple. She tossed her cigarette and studied the regular sized door embedded in the cinder block wall, anticipating her next move.

When the man and woman finally left, Pam unlocked the car hatchback. She lifted the heavy bolt cutter, grunted as she cut the padlock, and threw open the door. The putrid smell of rotten garbage made her gag. Covering her mouth and nose with her hand, she struggled to control her urge to vomit. After a few nauseating seconds, she succeeded.

Her flashlight cut through the interior gloom. The yellow beam swung onto the maggot-infested refuse in the rear corner of the room. From under the mess, the brass-tipped corner of a green trunk reflected the weak illumination. She hurried to the container, and used the bolt cutter to rake the trash to one side.

Pam crunched the trunk lock, tossed it aside, and yanked up the lid. She gasped at her find. Gold chunks filled the box. It was just as she had hoped. The guaranteed wealth staggered her imagination. "Yes, yes, yes!" she whispered. "Now, I alone, can control my destiny. To Hell with the many risks in Nolen's plan! No more dubious dreams. I'll be away from here in minutes. No one will know where to find me."

After concealing two large chunks of gold under her jacket, she carried them to the doorway. Glancing to her right and left, she verified that the few people at the far end of the storage alley were not watching her. Quickly, she strode to her car and dropped in her load.

25

Punishment

Jake Samson's 265 pound, six-foot six-inch frame consumed the car passenger seat beside Nolen. Jake dangled his long, muscular arm down to the floor. He lightly touched the sawed-off shotgun tucked beside his seat. His crew-cut head rotated from side to side, constantly checking the area around the moving auto. His movements would momentarily pause, when he spotted a car or person of interest. He would study that object, gauge its potential as a threat, then resume scanning.

Two weeks earlier the former semi-professional football player resigned as Susanville's deputy sheriff, to accept the position of Security Chief for the Association. Now, as he surveyed the entrance to the Blue and Orange Storage Company, he felt highly alert. He enjoyed his day's work, ensuring that Nolen safely retrieved his trunk of gold, and was not prevented from delivering it to Susanville. His new job was not boring, like the monotonous routine followed by a small town deputy. Every day had become exciting and different since Nolen had hired him. It revived some of the thrill he previously felt, before his knee injury forced him to stop playing football.

Rounding the corner to the line of twenty large storage rooms, Nolen slammed on the brakes. "Shit!" he exclaimed, then quickly backed the car out of view of the alley.

"What the hell's the problem?" Jake grumbled.

"Someone is parked in front of my bay. The door is open." Nolen jumped from the car and ran to the edge of the metal building. Jake slid to a stop beside him. Both men peered around the corner, studying the parked automobile 100 feet away.

"Look, Pam is stealing my gold. She is a barracuda!" Nolen stepped into the driveway.

Like a snake striking, Jake's arm shot forward, seizing his partner's collar. He yanked Nolen behind the end of the building. "Where do you think you're going?"

"Down there, to wring her double-crossing neck. That witch is betraying us. Now let me go!" Nolen's face was red with fury.

"It's too public! What happens if she has a gun and shoots you? The rest of us in the Association go down the tubes." Jake shook his partner. "Do you understand? This isn't the time or place to confront her. Keep your wits."

Nolen blinked, as he considered Jake's logic. "OK...OK, you're right. You've earned your pay for today."

Jake scrutinized Nolen, ensuring he had regained control of his emotions, then released him. Returning to the edge of the structure, the men watched Pam finish loading the gold.

They looked at each other. "We'll follow her," Nolen said, turning toward the auto.

"I'll drive," Jake ordered. I've tailed people before."

"You mean, you'll stay calm, when I might explode."

"Now you're rational again," Jake said, as he slid behind the steering wheel. He nosed the car into the alleyway. Pam's red sports car had disappeared. He sped between the buildings, and braked hard before reaching the far end. They spotted the reporter driving through the exit to the storage compound.

"Your job," Jake said, "is to watch her vehicle 100 percent of the time. Don't lose her. I may take my eyes off her car to weave around traffic. If she changes course, you tell me which direction, and how far ahead. Tell me every time she turns, whether I'm looking at her or not. Got it?"

"I understand."

The two men began their routine. Nolen echoed Pam's stops and turns, while Jake jockeyed their position through the traffic. He was careful to avoid being seen or approaching too closely. Once Pam merged onto the cross-town highway, their trailing became less difficult. She sped westward over the Bay Bridge, south through San Francisco, then west again toward the ocean.

Nolen spoke. "I think she's headed to her home, down the coast. If I'm right, in a mile, she'll take the Half Moon Bay exit ramp."

"Have you ever been to her place?"

"No, I just remember when she signed her Association contract, she listed her residence near Half Moon Bay."

"That's a big help, since it will be dark in another fifteen minutes, and we might lose her. Does she have a dog?"

"I don't know. She's single and travels a lot, so it's unlikely." Conversation ceased between the two men, as they waited for Pam's next move.

Her right taillight began flashing, as she approached the turn Nolen had described.

"She'll be alone at the house," Jake stated.

"How do you know that?"

"No one covered her at the storage bay. If she felt she needed help, that's where she would have used it. So, she's alone. Probably because she doesn't want to share the gold with anyone. She's a greedy one."

"Unfortunately, I think you're right. That's one vice of hers strong enough to ruin our plans!"

Jake switched on the headlights. "What do you want to do?"

"I want the gold back. And I want her punished for double-crossing us! On the other hand, she's crucial to the success of everything we've planned. To fuel the gold rush, we must have wide TV coverage. It's vital to rapidly spread the news of the gold discovery. There's not enough time to replace her. But I'm worried that, in the future, she might do something damaging like this again. She could easily release our story too early. That would cost us millions. Got any suggestions?"

Jake stopped the car. They watched Pam turn into a driveway, which stretched almost 100 yards from the road, down to a two-story house. The residence perched alone on a small, grassy promontory at the edge of the beach. Tall hedges protected it on two sides from the shore winds.

The big man bit his lip, concentrating. "How about a nice guy, bad guy drill. We create two types of motivations in her. Fear, she'll be hurt if she damages the Association in any way. And positive feelings toward you to keep her from being vindictive."

The automatic garage door rolled shut behind Pam, as she jumped from her sports car, clutching a large chunk of gold. She ran upstairs to her bedroom. There, she laid the rock on her dresser. She stared at it. Ever since she had opened the trunk, she had difficulty taking her eyes away from the fascinating ruby-yellow rocks.

She shook herself free from the metal's lure and walked to her closet. Selecting two expensive brown suitcases, she spread them open on her bed. In a frenzy, she flipped through her clothes, discarding unneeded items onto the floor. An armload of dresses, pants, and shoes landed beside the largest suitcase. From her dresser, she scooped lingerie, then piled them onto the bed.

Hastily, she stripped out of her garments. Her nose wrinkled at the musky smell emanating from her armpits. Again, the chunk of gold caught her eye. Picking it up, she hugged it against her chest. Her laughter echoed throughout the empty house. Kissing the lump of gold, she returned it to the dresser, then danced into the bathroom, humming.

Jake sneaked along the beach and onto the backyard grass. Light, streaming down from the second floor window, helped illuminate the patio area. The big man avoided the darker shapes of outdoor furniture, as he approached the rear of the house. When he reached the door, he ran his hand across the nine-pane window, embedded in the old wooden entrance. Stupid builders, he thought. They persist in selling customers attractive antique doors that burglars can easily penetrate.

Jake used his pocket knife. He cut away the putty caulking and wood which surrounded one of the small panes. Then, he pushed against the bottom edge of the glass, while prying at its top with his knife point. The pane quietly popped away from its frame, allowing Jake to snake his hand through the narrow hole to release the interior door latch.

Seconds later, he slipped into the dark kitchen, gently closing the door behind him. His senses probed the house. No sounds downstairs. No sign of a dog. He heard only a shower splashing, somewhere upstairs. He padded into the front room, then crept up the stairs. Jake quickened his steps as he stalked down the hallway and through the lighted bedroom doorway. There, he side-stepped, putting the wall behind him, just in case someone was in the room. Pausing, he noted the clothes strewn across the bed and floor. He heard the shower spray stop, followed by a door sliding, and a feminine voice singing. Jake moved toward the partially open bathroom door, absorbed in the adrenaline rush of the moment.

Pam was bent over, toweling her legs dry when she sensed a breeze cooling her back. She turned to close the door and jumped with fright. "Wha'...wha'...what are you doing here?"

She pressed herself hard against the shower door, cowering behind her towel. "Please...please...my money and jewelry are on the dresser. Take it all. I won't do anything to stop you. Please don't harm me."

"I'm not a robber," Jake snarled. "Your damn money is worthless to me."

A tremble shook the reporter's body, as she watched the huge man step closer and closer. He ripped the towel from her grasp. Using her arms and hands, she tried hiding her breasts and crotch from his view. "Please...please don't rape me!" she pleaded.

He grabbed her by the shoulders, yanking her away from the shower, then threw her through the doorway. She stumbled and fell to the bedroom floor. "You made three mistakes, bitch. Now, you're going to pay for them. Stand up!" he yelled.

She rose, with a shaking weakness in her legs. Her wide eyes assessed the danger confronting her. "I don't understand. What are you talking about?"

Jake swiftly backhanded the small newswoman. The force of the unexpected slap propelled Pam against the edge of the bed. She thudded to a seated position on the floor, dazed, feeling a shattering, hot stinging across the side of her face. A trickle of blood oozed from a cut on her swelling lower lip.

Jake knelt down and jerked her face upward. She was forced to look at him. "Your first mistake was not knowing Nolen had me watching the storage firm, to verify when it was safe for him to return. While you were loading the gold, I reached him on his car phone. He'll be here any moment. Now get up!"

She shook her head. "Please don't hurt me—please!"

He snatched her upright, and drove his fist into her stomach. Again, she crumpled to the floor, gasping for breath, withering in pain. He slapped and kicked her again and again, then stepped back and waited. She lay naked on the floor, vomiting and coughing blood. Red abrasions marked her ivory skin where she had been assaulted. She began crawling away from his wrath.

"Your second mistake was stealing the gold." He reached down, curled one hand into her hair, and shoved his other hand between her legs. She screamed with fear when he lifted her off the floor, effortlessly raising her above his head. Stepping forward, he flung her against the wall. She hit with a dull smack, then dropped into a heap. Blood gushed from her nose; tears streamed down her cheeks. Weakly, she rolled onto her stomach. Again she tried crawling away.

Jake stepped on her left hand, knelt down and grabbed her forefinger.

"No, please, I can't take any more. I'll do anything you want. Please! Please!"

"Your third mistake was not knowing I'm the sixth partner in the Association. Nolen didn't mention me at the meeting in Red Bluff. So, that's my gold you're trying to steal. Had you gotten away with it, I would have been screwed. So now...now, I'm going to kill you." Jake ripped her finger backward against the joint.

Nolen heard Pam's howl. He rushed up the stairs and burst into the bedroom. Jake sat astride Pam, strangling her. The reporter's face was beet red. Hell, maybe he's not acting, Nolen worried. Then he dove into the rib cage of his Security Chief, knocking him off the small woman. Scrambling to his feet, Nolen interlocked his fingers and slammed his forearms against Jake's chest.

"Stop!"

The glaze in the big man's eyes faded.

"Clear off the bed, and get me a wet towel," Nolen ordered. He turned and picked up Pam.

Jake returned from the bathroom and threw the cold cloth onto the woman's stomach. She jerked when it hit her.

"Go downstairs and wait," Nolen directed.

Jake spun around. "Dirty bitch!" he swore over his shoulder, as he left the room.

She clung to Nolen, fearful, crying and trembling. Blood, mucus, and tears soaked his shirt. "That animal tried to kill me! Please, send him away. I'm sorry...so sorry."

As she cried, Nolen cleaned her face, pulling strands of hair from her mouth and wiping the blood away. She winced, as he touched her swollen cheeks and dark bruises tattooing her chest and arms. "Why did you steal the gold?" he questioned.

"After the meeting, I was afraid. I just wanted reassurance we had enough money to finance your plan. I remembered seeing another trunk in the storage room." Pam sobbed, gasping for air, then calmed, and resumed answering. "I planned to take the gold back to Susanville. I was going to make you give it to Hamilton. Don't you see, I was trying to protect you!" Tears streamed down her cheeks.

"You could have confronted me about the second trunk," Nolen countered, "in front of the rest of the partners. That would have forced me to bring it in. Those suitcases you were packing reveal your lie. Don't bullshit me, Pam. Admit it, we would never have seen you again."

"I'm sorry. I know I shouldn't have done it."

He shook her. "Listen! The only thing keeping you alive is me. If the other Association members learn what you attempted, I won't be able to stop them from killing you. I'll control Jake. He'll keep his mouth shut, only because you have valuable work to do. But from now on, a guard will be with you, day and night. Should you try anything dumb like this again, I won't be able to stop Jake. Furthermore, I won't even try."

"Nolen, I promise to do exactly as you say. I promise!"

He unwrapped her clinging arms and helped her stand. "Take a shower, and get dressed. We need a doctor to mend your hand."

Nolen left Pam and found his Security Chief standing in the kitchen, sipping a beer. Nolen accepted the unopened can offered to him. "It appears, Jake, our charade worked. You scared the living daylights out of her."

"Yup, I did. She'll think twice before crossing us again."

"All right, Jake, I'll collect the gold and leave for Susanville. Have a doctor examine her hand and check if she has any cracked ribs. Explain she was in a car accident. Bash in the front of her sports car. That will substantiate our story. Then, get a couple of guards to stay with her, until we file the claim. However, don't frighten her any more. Understand?"

The big man crushed his beer can, in one hand, and tossed it into the sink. "Sure, Boss." He headed upstairs.

Nolen took a deep breath. Shaking his head, he thought, I set out to do something worthwhile, something Digger dreamed about during his years of prospecting. I added logical steps to pull off his plan, additional moves and counter moves. But now I'm hurting people, deceiving some, tempting others. It's nearly impossible to recognize and then defuse the damnable emotional needs of everyone before they explode. All the others are becoming scared too, getting close to jumping into the personal survival mode. Maybe I can pacify their anxiety for a little while longer. But what if I can't keep them in control?

26

Cornered

12 September

The young FBI agent savored the fragrant fresh morning air, as he hurried up the steps leading into the quaint, small-town library. Inside, shafts of sunlight brightened the quiet interior, highlighting the specks of dust floating in the air. A thin, older woman fumbled with a pile of books stacked on a push-cart.

"Hello. Are you the branch librarian?"

The arthritic matron looked up at the athletic young man. Her eyes flickered behind thick eyeglasses. "Yes, I am. What can I do for you?"

He smiled. "My name is Cholo Cantera. A clerk at the courthouse referred me to you. She said you have been a resident of Susanville for thirty-five years."

"That's right, young man. I moved here with my husband in 1965. What subject are you researching?"

"I'm interested to know what Federal projects have been active in this county, during the past twenty years. Since you are a resident, you are the perfect candidate to tell me how this county has been influenced by those projects."

A slight smile creased her wrinkled face, as she carefully sat on a nearby padded bench, and modestly adjusted her hemline. "Young man, I have an easy answer for you. There have been only a few state and federal projects. I think I can describe all the major events. Of course, you can also browse through our reference files. The greatest impact came from the depot. You know about it, don't you? It's out by the border, east of town."

"Yes ma'm, I noticed it on my road map."

"It's quiet now. During the war, several thousand people were employed there."

"Which war?"

"Oh, how silly of me. There have been several, haven't there? I'm speaking of the Vietnam war. Those were the days when every family in town had at least one relative working at the base. However, since about 1982, the number has dwindled, to only fifty or so locals. Most of them relocated near the base. They seem to keep to themselves more now, not very friendly."

"Time changes people and towns, ma'm," Cantera interjected.

She continued. "I think it was, in '91, they did some considerable construction there. Folks were hoping the depot would be re-activated because of the fighting in Iraq. If my memory serves me correctly, numerous newspaper editorials were published in the fall of that year. They encouraged voters to write their congressmen to request that the depot expand its operations. It didn't happen though. Let me show you the available reference material."

Cantera sat in front of a microfiche reader. He recalled his favorite academy instructor teaching him about the usefulness of investigative research. "Many times, your best information is hidden within the mundane pages of old reports, ledgers, notebooks, diaries, newspapers, or file folders."

He began reading back issues of the local newspaper, as well as county records. It was a slow, tedious task, since he had to work his way through 106 Photostat rolls, preserving two decades of information. Repeatedly, he threaded the tape into the projector, then began reeling. Articles regarding federal agencies and mining were his main focus. When he spotted a headline of interest, he scanned the first paragraph to determine if the article might contain useful information.

Time after time, he found nothing of value. Gradually, a pattern evolved. Cantera discovered that three federal agencies were located in the valley: the Army, the National Forest Service, and the Bureau of Land Management. He further learned that the last successful prospecting in the county had occurred in the late 1920's, when a small gold vein was uncovered. The librarian had been correct. The most recent military construction happened in 1991 through 1992. One non-routine story surfaced, describing how the Forest Service responded to a recent problem with herbicide poisoning in the local forest.

Hours disappeared. Cantera grew weary. His shoulders ached. An irritating blister stung his right hand, formed from hours spent cranking the projector handle. Rocking back in his chair, he stretched and considered his poor progress. I'm certainly seeing the mundane pages, but no gems of information. Perhaps, I've wasted my time. Hopefully, the police records will generate something more substantial than these skimpy stories. After tossing his notebook into his briefcase, he left the building.

Across town at police headquarters, Cantera reviewed the case file on Digger Jorgenson's traffic accident, as well as the report detailing the gas station murder. He ceased speed reading when he reached the third page of the homicide investigation. Well, well, well, he thought. The murder victim's car was found parked, down the road from the scene of the crime. That may indicate the man followed someone, then confronted him at the gas station. Papers in the car proved the dead man worked for Continental Mining and Refining. Yet, CMR claimed no knowledge of why the dead man was at that station.

The agent dug one of several note pads from his attaché case, then thumbed through it. He read an entry gathered by the Reno wiretap. Boodan had stated he was trying to frame Martin, by having some woman claim she witnessed Martin leave the station. Cantera smiled. It's not a gem, by God, but the police file, combined with the wire tap information does seem to corroborate Martin's innocence of the killing. CMR's man probably tried to capture Martin at the gas station, but got himself killed in the attempt. Well, one more source to verify, then I should be through with this town. He returned the police files. Directions to the only hospital in town were his final request.

A nurse led Cantera to an office on the second floor of the hospital complex. In the room, a man sat at his desk with his back facing the door, reviewing a medical journal. "Doctor, you have a visitor," the nurse announced.

"Yeah," he said inattentively. "I'll be right there."

"Doctor, the gentleman is here with me."

The physician turned his head, surprised to see two people standing in the doorway. A frown wrinkled his features, because he did not recognize Cantera. The nurse smiled at Cantera before she left down the corridor.

"Doctor, I'm special agent Cantera, Federal Bureau of Investigation." He displayed his identification card and badge. "I need to question you regarding a case you were involved with six weeks ago."

The older man raised his eyebrows. "Which one?"

"Hospital records indicate you were the attending physician to a patient named Digger Jorgenson. He was admitted following an auto accident. Can you supply more details concerning that patient?"

"Sure. I remember that case quite well. Why is the FBI interested in Jorgenson?"

"We believe his accident may be linked to a murder in this area, which occurred only a few weeks later."

"Since we don't have many slayings here, you must mean the one at the gas station."

"You're right. Can you help me?"

"Easily." The doctor swiveled around to his desk, opened a drawer, and extracted several sheets of notes from a file. "I'll keep most of the medical jargon out of this, as I proceed."

"That will be helpful."

"Jorgenson, a Caucasian male, age sixty-seven, six-feet two-inches tall, 180 pounds, was transported to the hospital Emergency Room at 2:16 a.m. The police said they extracted him from his truck, which had veered off the road into a ditch.

"Initial prognosis: The patient was unconscious. He had suffered several injuries during the accident including: left lower leg fracture, sternum fracture, and head concussion. While in the Emergency Room, he experienced a myocardial infarction. Excuse me, that means a major heart attack."

"I am familiar with the medical term."

"By the way, the Sheriff's Department did a great job of getting him to the hospital quickly. A bartender, closing his tavern, witnessed Jorgenson weave along the highway, then slam into the ditch. Luckily, Jorgenson's accident occurred on the edge of town. Not that it did him much good in the long run. But the heart attack would have been fatal if he hadn't been downstairs in ER at the time.

"Anyway, once the patient's condition stabilized, he was transferred into our Intensive Care Ward. Ten hours later, he lapsed into a coma. He was alert just long enough for his family to speak with him, before he slipped away."

Cantera stopped writing in his notebook to look at the doctor. "Your information is helpful. Do you maintain such accurate records on all your patients?"

"Well, this case was special."

"Why is that?" Cantera inquired.

"The day before Mr. Jorgenson's accident, there was a herbicide poisoning scare, caused by aerial spraying in the nearby national park. Consequently, we were on alert for any arriving patients, who might have been affected. The Forest Service was concerned that the wind might have carried chemicals onto some of the locals. They even called in a virologist, Dr. Jeffery Daly. He flew here from the Atlanta Center for Disease Control, to help us screen patients."

Adrenaline shocked the agent. Had he finally uncovered the type of revelation he was seeking?

The physician continued. "Most Californians are real ecology supporters. Such incidents usually lead to a big uproar in this state. That's another reason for maintaining detailed notes on patients handled during that period."

"Was anyone hospitalized with poison symptoms?"

"No. Dr. Daly checked everyone who entered the hospital during his three-day stay. It was a pleasure working with him. A true professional, most proficient. When Jorgenson was admitted, Daly insisted we immediately examine for poisoning."

"Wouldn't you have done that anyway?"

"Yes. But it was Dr. Daly's conscientious attitude that impressed me. He wasn't afraid to don his greens to assist us. He helped with the lab work, ran the toxicology analysis, and most of the biopsy tests too."

"You mean those tests were ordered on Jorgenson?"

"Yes."

"And, I take it, there were no signs of herbicides in his system."

"No, none at all. All data indicated Jorgenson's heart attack and coma were most likely induced by the auto accident."

"I'd like to obtain a copy of your notes, as well as all official reports, if I may?"

"Sure, no problem," the doctor replied.

"Would you have a phone number or office address where I can reach Dr. Daly?"

"I recall his letter of introduction, sent by the Forest Service, listing Dr. Daly's qualifications. It also included details about the CDC office which he was representing. The letter is filed downstairs. Come with me to Administration. We'll photocopy everything you need."

Minutes later, a clerk returned from her search for the documents. "I'm sorry, Doctor. I distinctly recall that letter and file, but I can't seem to locate them. I made a thorough check. It might be misfiled."

Cantera wondered, Perhaps it's just coincidence that those records are missing. Or, she'll never find the letter, because the file has been destroyed. He concluded his investigation by acquiring a description of the CDC doctor. Then he left the hospital, and drove back to the library. There, he copied the articles explaining the herbicide poisoning incident.

13 September

Early the next morning, Cantera contacted the Center for Disease Control. He was not surprised when he could not locate Dr. Jeffery Daly as a permanent staff member. Nor was Daly a standby medical expert from any affiliated hospital in the country. Furthermore, the CDC representative stressed the fact that they had never sent anyone to Susanville to screen patients for possible herbicide poisoning.

Who the hell does this doctor work for? Cantera asked himself. Well, I'll check with the Army and the Forest Service. He made phone inquiries to the Army's personnel records center in St. Louis and to the Forest Service's regional center in San Francisco. Late in the day, he received a return call from Missouri.

"Sir, this is Master Sergeant Cooper, from the U.S. Army Personnel Data Center, returning your call. I think I may have something for you."

"Great, Sergeant, what have you learned?"

"There is an Army medical officer, a Lieutenant Colonel Jefferson K. Daly, assigned to Brooke Army Hospital at Fort Sam Houston, Texas. He's an epidemiologist. The computer search did not reveal any other individual, meeting the characteristic variables you requested."

"What does an Army epidemiologist do?" Cantera questioned.

"He studies what causes infections, that type of work. I have his service record available. Should I fax a copy to you today?"

"Yes, Sergeant, I'm very interested in your findings. Your staff has done a superb job. Can you tell me, was he ever stationed in California? Has he worked with epidemics, or perhaps chemical poisoning incidents?"

"Hold on." The phone went silent. "Got it. No tour in California. His nearest assignment was Dugway Proving grounds in Utah. The only indication of chemical assignments is his work at Fort McClellan, Georgia, on two different occasions."

"Fort McClellan?"

"Sorry, that's the Army's Nuclear, Biological, and Chemical Center and School, sir. He may have done some work there on chemical poisoning. While assigned to that post, he was the Chief of the Virology Lab. But, I'd say from reviewing his record, his expertise lies in the biological field, not chemical."

"Thanks, Sergeant. I appreciate your efforts." Cantera began analyzing the new information. Well, the SOB's have sure screwed up this time. I can prove the Army's Lieutenant Colonel secretly tried to determine if Jorgenson was poisoned by some unknown compound. Jorgenson is Martin's foster father. And the FBI is searching for Martin. It's a direct chain, connecting the FBI to Martin, to Jorgenson, to the Army, to a possible poisoning.

If there is a link, some very powerful people could refute my claim, with suddenly found records. They could show Daly was in Africa or the Antarctic, at the time of the incident. I need more proof to protect myself, as well as to prove my case. I'll make a couple of calls to the CID and SEC. Then, I'll flush the quail from the bush. And, hopefully, I'll learn what the government is hiding.

14 September

At eight o'clock the next morning, Cantera entered the large stock brokerage firm, situated on the third floor of a corporate building in downtown Oakland. More than fifty men and women were working in the cavernous room. Most of them were busy conversing. Others hurried from desk to desk. Several typed at computer terminals.

The agent informed the receptionist he wished to speak with Miss Collins, concerning purchasing some stock.

Moments later, Maida politely smiled as she stepped forward and shook hands with her new client.

Watching Maida's reaction, Cantera inquired, "Can you help me acquire some Benjamin Mining stock?"

Concern flickered across Maida's face, upon hearing a stranger request the stock she had secretly been buying over the last several weeks. "Let's step into a private room," she suggested.

Inside the room, Cantera waited until the broker closed the door, then he displayed his identification. "Miss Collins, I'm with the Federal Bureau of Investigation."

The color drained from Maida's complexion, as she slipped into a chair. "I don't understand. Why are you talking to me?"

"Don't try to mislead me," the agent warned. "I have been assigned to track Nolen Martin. My efforts have uncovered enough evidence to arrest you. I know Martin discovered a gold strike several weeks ago. He is involved in the death, possible murder, of a man in Lassen county. And FBI wire taps prove Continental Mining and Refining is seeking to separate Martin from his mineral strike."

"Slow down," Maida urged. "You should be contacting CMR, not me."

"Miss Collins, you are in trouble. I had the Security and Exchange Commission analyze your recent stock transactions. You have been active in purchasing mining stock and gold commodities for Nolen Martin."

"You can't prove that," Maida argued.

"Screening by the SEC revealed no Benjamin Mining stock transactions during the past twelve months, until your first purchase. Shortly thereafter, business telephone records verify you called five West Coast offices. Those calls occurred just prior to initial purchases of Benjamin Mining stock by a broker in each of those offices. Now, you and the five other brokers are buying stock at a rate which violates the Hart-Scott-Rodino Act, requiring disclosure of takeover attempts."

Maida struggled to hold back the tears welling in her eyes as the agent droned on.

"Due to your direct involvement, you can be charged as an accessory to murder of the man Martin killed. I can put you in jail right now and then go public with the news of Martin's gold discovery!" Cantera paused, providing Maida ample opportunity to worry.

She closed her eyes, clenched her fists, then asked, "What do you want from me?"

"Explain why you are involved with Martin. Tell me why you are buying Benjamin Mining Stock. Then, I want to meet with Martin."

27

Trust

Ghirardelli Square bustled with midday activity. San Francisco's personality was on display and for sale. Painted clowns selling glittery, silver balloons performed fanciful magic tricks. They easily captured the amazed attention of young children, who gazed wide-eyed as multicolored scarves appeared, then disappeared.

Competition for the tourist money beckoned only a few steps down the sidewalk. A talented black couple, dressed in white formal wear, top hat, gloves, and brandishing straight canes, responded to music resonating from a cassette tape recorder. In unison, they tap-danced for the crowd. Across the street on the grass mall, a throng of people watched a pretty girl balance on a unicycle, while juggling three balls of different sizes. Lovers, families, street people, and Asian businessmen streamed into, up, down, and around the magnificent red brick emporium.

Within the complex, Nolen and the FBI agent approached the main fountain from opposite directions. Dozens of visitors jostled by, talking and laughing, as the two men scrutinized one another. Both men kept their hands visible to ensure neither had a gun trained on the other. They sat on an unoccupied bench, ignoring the bustle around them.

"I'm here as Maida requested, Agent Cantera. What is it that you want?"

"For starters, I need the facts about your foster father's auto accident, and his subsequent hospitalization in Susanville. Also, what have your actions to do with the Army depot in Lassen county? Then, give me details as to how the CMR man was killed in the gas station."

"Why should I explain anything to you?" Nolen responded, somewhat hostile.

"You don't have much choice. If you don't cooperate, I will inform the news media about your large gold strike, together with its general location. I venture, such news would ruin your plans. Or I can arrest you on suspicion of murder. Or, I can arrest you for violating SEC regulations. Take your pick."

"I have more alternatives than you think," Nolen replied. "Even if I can't conclude my plans, I know the precise location of the strike. I can retrieve millions of dollars worth of gold any time I want. So, I can walk away from all my preparations whenever I choose. Anyway, I won't allow you to arrest me."

The agent recognized the implications of Nolen's last statement, and glanced around the square, feeling his buttocks tighten as he considered his vulnerability. Hell, he's right! I'm here without backup. I could get creamed. My best alternative is to bluff my way through.

Cantera replied. "A shoot-out between your men and mine won't do either of us any good. Furthermore, if I'm killed or disappear for some unknown reason, then the report I've written about your secret will automatically be made public. Why don't we exchange information?"

This time, Nolen slowly scanned the square. "You're right. We have more to gain by sharing major bits of information. I'll volunteer some facts I think you need. Then, you provide some information I need. That way, each of us has to take some risk in exposing what we know."

The young agent nodded. "Rather like wandering through a minefield, with each of us holding half the map. Go ahead, you take the first step."

"I think these facts are pertinent. On August 14, while I was prospecting, a 737 jet flew over my campsite. It woke me. The plane then landed inside the Army operated Sierra Mountain Ammunition Depot. The plane landed on an airstrip that doesn't appear on any U.S. Geologic Survey maps. I watched as it was loaded in the dark with some cargo. The aircraft remained on the ground for less than a half hour."

"What time did the plane fly over your camp?"

"About 3:00 a.m. The next day, after I was almost killed by a flash flood, I left the site. As I drove from the area, I found two men who had not been so lucky. Both were dead. Each man wore what was left of a U.S. Forest Service uniform. But I found an ID card on one, indicating he was an Army E-7, named Thurman. I also located their vehicle, which had an interior mounted encryption device attached to a military radio. When I saw that secure military equipment, I knew the Forest Service vehicle and uniforms were a cover. That's when I got the hell out of there. Now, it's your turn."

The agent considered the implications as several parts of his puzzle fell into place. "There's something secret transpiring at the depot. Ever since you saw the flight and the dead sergeants, the government has been using the FBI to track you down. The military must have some connection with my boss, the FBI Regional Director. I believe he created a false intelligence file, portraying your involvement with a right-wing radical group. That file alleges you stole weapons from the Army."

"You bet it's false," Nolen snapped. "Just review my military record. It will attest to my loyalty to my country."

Cantera continued. "Linking you with that terrorist robbery gave the FBI a valid motive to search for you. It was also
a ploy to mislead any agents chosen to find and arrest you.
My guess is, my boss selected me for this assignment, because he thinks I'm too inexperienced to realize the government wants you for other reasons.

"But, during my investigation, I learned that the Hunter
Liggett soldier who was attacked never saw his assailants.
That fact, coupled with a few others, finally convinced me
that you are not involved with any radicals."

Cantera looked about the small plaza, verifying no one
suspicious had approached his back, then continued. "A
specialist was sent to the Susanville hospital. Supposedly,
the National Forest Service requested help from the Center
for Disease Control. His job was to help identify anyone who
may have been poisoned by an accidental herbicide spraying, in the local forest. He observed your father, when he
was admitted to ER, and later assisted on the subsequent pathology analysis.

"Turns out, that specialist is actually an Army epidemiologist, a trained expert on biological agents. This guy has
never had any affiliation with the CDC. My theory is—the
military is transporting biological agents to or from that depot. They had an accident about the time your father was
prospecting. And the Forest Service-CDC concept was a perfect cover-up for the leak at the depot."

It took all of Nolen's willpower to contain the rage surg-
ing through his body, as he recalled his father's injuries. In-
ternally, he swore at himself. *I should have realized
something was wrong when I saw Digger's unnatural, pallid
complexion.*

Nolen furrowed his brow and rubbed his chin, attempting to conceal his anger. "That explains why you've been
tracking me, all right. But that stuff about a leak isn't much
help. Digger just seemed weak from his heart attack and

other injuries, when he lapsed into the coma. I didn't recognize any signs of poisoning."

"Well, that is good news. Are you aware of a nerve gas accident the Army experienced in the mid-sixties? That event killed thousands of sheep on Dugway Proving Grounds in Utah. Luckily, no humans were hurt that time. And God bless us, apparently no one was hurt this time either."

Nolen informed the agent how he had been followed from the assay office, attacked by the two CMR men, and later formed the gold Association. Then, he asked, "How did you learn about Maida?"

Cantera detailed CMR's burglary of Nolen's home. He explained how the phone tap in Reno revealed CMR had contacted Nolen's ex-wife, and his foster mother, for information. Cantera further described the attempts by the exploration company to illegally gain control of Nolen's discovery. He elaborated on how CMR knew about Maida, and the subsequent FBI tap on Maida's telephones.

Nolen was puzzled. "I can't quite understand your motives. You know I have violated some SEC regulations. Yet, you're readily providing information which benefits me a great deal."

"The government and CMR are committing much greater crimes than you, Martin. The Army is probably violating a 1969 Presidential order, which forbids anyone in the USA from developing biological warfare agents. Without adequate safeguards, the Army has exposed the public to some very toxic material. Meanwhile, CMR appears willing to commit kidnapping, burglary, conspiracy, and fraud to reach their goals. Your illegal take-over maneuver is minor by comparison to what the government and CMR are up to."

"That's how I've felt all along," Nolen agreed. "You know, however, what you're doing may ruin your career?"

"I don't need a career that converts me into a secret police agent. I was raised to believe in our justice system, to be a protector. Someone who could be trusted to fairly enforce the laws of this country. Perhaps, I'm too idealistic, but I intend to do just that. You might say, I'm ensuring the scales of justice are not unfairly weighted against you, by your own government."

"I appreciate your honesty," Nolen interjected.

"Don't think for a moment that my feelings allow you free rein to do anything you please. My options, like yours, are still open. Realize, I can always arrest you or Collins before or after you file your claim. However that won't happen, if you cooperate. I need your assistance to verify whether or not the government is involved in something improper. So I'm making you an offer. Don't break any more laws. Let me know where I can contact you when I need information. Then, I won't interfere with your filing." Cantera handed a card to Nolen. "Here's my number."

Nolen felt respect for the principled agent seated next to him. He also felt great relief at not having all his progress smashed. "You have a deal, Cantera. Now I'm going to take a big chance." He scribbled an area code and seven digits on a piece of paper and offered it to the agent. "Call this number any time of the day. I or one of my people will then respond within an hour."

As Nolen walked away, his Security Chief emerged from a store where he had waited, observing the meeting. He fell into step beside Nolen. "What's the matter?"

"Jake, we've got to move very fast. I want bodyguards protecting Maida, immediately. Also, have someone check if my foster father's crashed truck is still in the junkyard in Susanville. While you're handling those tasks, I must speak with my foster mother. We're in danger of losing everything."

28

Construction

Boodan had flown over the desert construction site two days earlier. He wanted to view for himself the mysterious town springing up along the state border. As his small plane circled, he had studied the recently built unloading ramps, stacks of supplies, as well as the many vehicles, moving about the partially constructed structures dotting the sand. Boodan maintained it was not coincidence that within the same county, two uncommon events had occurred, the discovery of a large gold strike and the emergence of a new town. Boodan suspected that Martin was involved with the construction and might appear at any time. That act would give Boodan an opportunity to snatch him.

Now, the Cajun inched his pickup forward, trapped in a line of vehicles. He removed his baseball cap, wiped sweat from his brow, then resettled his cap, watching the afternoon activity surrounding him. Six men, wearing bright orange vests, were split into two teams of three. They controlled the traffic entering and exiting the four-lane entrance to the immense, fenced compound. One flagman, hoisting a stop sign, checked each approaching vehicle. He waved into the inner lane, all trucks with a red identification tag pasted to their windshield. They received immediate access. Gravel, cement, and loaded flatbed trucks roared past the vehicles held in the outer lane, waiting to enter.

A second man, clutching a clipboard, approached Boodan. "Where are you going?" he shouted, loud enough to be heard over the engine clamor.

Boodan grabbed his authorization sheet from his cluttered dashboard. Poking it toward the man he yelled, "The guy hirin' people in Susanville told me to report here."

"You were supposed to ride the worker's bus. I can't let you inside with your pickup, sorry."

"Hey, they hired me late this mornin', after the bus left. Said I had to replace another fella who didn't show." Boodan did not mention how he broke the real hiree's leg, nor that he stole the man's hiring sheet. "I was told to get out here right away. Look, I don't have nowhere to leave my truck. Ain't there someplace I can park?"

Behind Boodan, an anxious driver started incessantly sounding his horn. Several others joined in with enthusiasm. A panel truck rushed by, blasting dirt and debris over the man with the clipboard. Instinctively, he dropped his chin to his chest. Looking back along the line of impatient deliverymen, he yelled, "Cool it, buddy! I'll get to you."

He returned the authorization to Boodan. "Some assholes are always agitated. That guy on the horn might never get past this gate." Pointing toward the camp, he continued, "See the white trailer, over to your left? Report to the Site Supervisor. Just don't block any roads, or park near ground cordoned off with colored tape."

"Right." Boodan eased forward, then waited until the flagman gestured for him to pull through the entrance. He traveled 200 yards, then veered from the hardtop and onto a sandy road, recently scratched out of the sagebrush. After parking, he watched the many work crews scattered throughout the camp. Survey teams, using red, blue, and orange plastic tape, were implanting stakes to surround large patches of pale-yellow sand. A bulldozer lumbered about, leveling one such area. It shoved excess dirt and brush into a pile, where a scoop loader removed the material and dropped it into a waiting dump truck. Clouds of dust drifted onto nearby workers.

Boodan walked to the trailer where the Site Supervisor was arguing with another man. "Damn, this is the third time this week those cement trucks have caved in pipe trenches. That forces me to return the backhoe team to a site where we've already dug. Can't your drivers see those damn ditches? They don't give a shit! The next time I catch some guy driving like a dumbbell, I'm going to yank his ass out of his cab, and fire him on the spot. You tell them that!"

"OK, I'll tell my men. But how about changing the sequence? Let's dig the trenches after the cement trucks drop their loads, not before."

The supervisor reviewed his paperwork, then agreed. "All right. Go inside and coordinate the changes, so we don't keep falling further behind schedule." The construction supervisor turned to see Boodan approaching. "What do you want?"

"I was told to get my assignment here. I'm supposed to be a roustabout."

The harried supervisor smiled. "Great, I can sure use you. We haven't been able to hire people fast enough to fill the teams for every task. So, you'll be working alone for a day or two. Come over here." The two men stepped next to a table supporting several large maps, pinned to an oversized drafting table.

The supervisor unrolled a bulky blueprint. "Here's a diagram of the town. That building behind us is the Administration Headquarters. The two-story building down the way is the workers' quarters."

"Go ahead, I'm with ya," Boodan mumbled, memorizing the map.

"All right. These little X's indicate mobile toilet sites. The small squares are garbage trailers. Get a tractor from the warehouse, and emplace the johns and garbage trailers. You're a strong looking guy, so you shouldn't have much trouble handling this job by yourself. You'll be responsible for emptying the toilets and hauling the trailers to the refuse dump each day."

"I'm the garbage collector?"

"You got it right. And there's plenty for you to move. But first, get a bunk at the workers' quarters. Then, go to the California warehouse. Sign's on the front." The supervisor scribbled an authorization across a piece of paper, which he thrust at Boodan. "Give this note to the foreman from the storage yard. He'll issue your equipment. Any questions?"

"Nope. Just seems like a shitty job."

The supervisor chuckled. "Dry humor, I like that. Listen, things may look disorganized. It's just because we're busy as hell, trying to meet our tight deadline. We're really following a strict schedule. Every job is important, needs to be done on time, and done well. Inspectors are checking the assigned work, so don't goof off. Do a good job, ahead of schedule, and you'll earn a bonus. Waste time, and I'll dock your pay."

"That's what they told me when I signed on."

"We're running behind getting refuse receptacles distributed, so I need you to move quickly on this task. The head man's arriving tonight, and I don't want to irritate him because he can't find a comfortable spot to take a crap."

"Some big shot, huh?"

"He pays the wages, and it's his first visit."

"What's his name?" Boodan questioned, digging for more facts.

"Don't worry about him. It's me, and my inspectors, you have to keep happy. Report here, after your evening meal. See the Assistant Site Supervisor. Let him know what you've accomplished. Now get going." The manager abruptly ended the conversation. Turning away, he resumed studying the blueprint and a much-handled delivery schedule.

After hiking to the two-story worker's building, Boodan stepped into the wide entrance foyer. There were no chairs or benches, just one large stairway on the right, leading to the second floor. Two doors were built in the center of the wall, opening into the first floor quarters. To the left, behind a wooden counter, stood a small, wrinkled man, who looked up from a bin where he was working. "You new? You look new."

"Yup, just got in," Boodan replied.

The scrawny old man limped to the counter. "Name?"

"Boodan Tribou."

"Boo who?" The old man cackled momentarily, but abruptly stopped when he realized the menacing man before him did not share his humor. "OK, Tribou. You on day or night crew?"

"Huh?"

"Boy, you are new. Everyone here works 12-hour shifts. Ain't you been told? The night shift sleeps on the second floor. The day shift sleeps on the first floor. That way, things are quieter for the off-duty crew trying to get some rest. So, do you work when the sun is up, or when it's down?"

"Day shift, I guess."

The old man flipped to a page in his green ledger. "Sign here, then print your name so I can read it."

When Boodan finished the task, the old man spun the book around to read the entry. "OK, your bunk number is FB-156. That's first floor, bottom bunk, bed 156. Don't cause me no trouble by hassling the guy above ya."

"Right."

The scraggly-haired clerk suddenly disappeared below the counter, then reappeared offering two blankets and a pillow. He bent down again, this time returning with a laundry bag, sheets, and a pillow case. From a shelf, he selected three padlocks, which he laid on the counter. "Now, Tribou, you get two wall lockers and one foot locker to stow your gear. They are also numbered FB-156. You'll find them next to your bed. Keep your stuff locked up. We go through the building four times a day. Any crap left hanging around, or on the floor, gets thrown into the street. We're not responsible for any missing personal items."

"OK, Mom."

"Smart mouth, I got nearly 800 beds here, and I can't afford a mess, fights, or noisy guys who are a pain in the ass. My job's to keep it quiet here, so everyone can sleep. You'll be tired as hell a day or two from now. Then, you'll understand. So, mind my rules or sleep outside in the sand." The slight clerk shoved his head forward and squinted at Boodan. "Understand?"

"You don't have to say it twice, Killer."

"All right, if you got laundry, put it in your bag. Drop it off here Tuesdays, Fridays, and Sundays. I'll get it back to you. Any questions?"

"What about meals and a shower?"

"Chow for the work crew is served at kitchen number one. Out the door, turn left, down the road to the main intersection, and off to the right. Just follow the signs. Meal times are at 6:00 a.m. and 6:00 p.m. You'll be given a big sack lunch at breakfast. Use bath house number one to get clean. It's open twenty-four hours. It's near the dining hall. Any more questions?"

"No Mom, that's all I need to know."

"Good. Go through them center double doors and find your bed. It's at the far end of the main aisle."

After stowing his gear, Boodan located the warehouse, on the California side of the border. There he signed for his tractor. Earlier, upon hearing his new job description, Boodan realized his duties would allow him the freedom to travel throughout the town, during the day or night. Now, to gain an overview of the town, he drove through the bustling compound, towing his first string of wheeled garbage bins.

It was easy to recognize the construction plan. The northern end of town, containing the railhead, warehouses, and storage yards, had been developed first. Train cars rolled into that section of town with their loads, previously spray-painted, indicating for which building the cargo was intended. Similarly, semitrailer trucks would stop at the warehouses, where they were directed to the appropriate delivery points. Numerous orange forklifts scurried about, unloading material and equipment from the rail cars and supply trucks.

Moving south from the railhead, buildings appeared in different stages of construction. Frames and walls were erected for some, cement slabs poured for others, while only wooden foundation stakes outlined building sites at the southern end of the camp. Throughout the compound, men were fastening exterior siding to buildings. Others were painting, while fixtures, benches, or large boxes were being moved inside the structures. Piles of scrap building material and discarded debris littered and obstructed the entire area.

As Boodan observed the activity, a wicked idea came to him. He realized how he might slow Nolen's tight schedule, and how he might also snatch the miner during the ensuing confusion.

29

Fire

At dusk, a Piper Cherokee flew out of the southeast and circled the work site. For the first time, Nolen inspected the physical progress of his plan. Bright lights illuminated the men and machinery moving about the darkening desert. A crane hoisted a prefabricated wall into place, as workers swarmed around it, securing the metal partition to its cement foundation. From his vantage point, Nolen studied the gravel streets outlining the skeleton boundaries of the emerging town.

The small plane banked toward the runway. Minutes later, Nolen met his brother inside the hanger. The two men immediately boarded a jeep, to begin the bouncing ride toward the town. A second four-wheel drive truck, transporting the security team, escorted the jeep.

Bumping across the railroad tracks, they approached a side entrance in the fence. Nolen pointed to a red and white rectangular billboard. "What does it say?"

"Studio and Set Construction Incorporated. Authorized Personnel Only," Paul answered. "We needed a cover story to prevent people from scrutinizing us too closely. So, I devised the idea of a movie set. Simultaneously, Hamilton spread the word that a Hollywood company would begin filming within the county during the coming months."

"Has the ruse done us any good?"

"Since we hung the signs, the workers have quit speculating about why the town is being built. Yesterday, a newspaper reporter from Susanville came here asking questions. I gave him a brief interview. Told him my specialty is erecting sets on location for film companies. He inquired about the subject of the movie. I explained that I was rarely privy to the story lines. I just build what's drawn on the blueprints. Subsequently, we talked about the prospect of providing work for local residents. All the facts seemed to fall conveniently into place for him."

"Great idea, Paul!"

"You'll also be satisfied with the headquarters. To accommodate each of the partners, I provided a room, complete with a desk and a sleeping area. They're not spacious, but the rooms seem adequate. The front of the building houses a large conference room. I installed our communication complex adjacent to the security police offices, in the rear."

As Paul concluded his update, the jeep moved past a line of cement trucks and the headquarters building came into full view. Floodlights, secured to each corner of the structure, brightened the area surrounding the silver metal building, including a black limousine parked beside the doorway. A chauffeur was wiping dust from the polished vehicle, while two Association security men stood nearby, conversing as they puffed their cigarettes.

"Are those men guarding the headquarters or the car?" Nolen inquired.

"They watch the building," Paul replied. "Jake and I agreed the headquarters, warehouses, banks, and bordello all require individual guards. Further, we established three roving patrols, twenty-four hours per day. One team covers the area north of Main street, since that's where the majority of our inventory is stored."

"Fine. What about the limo?"

As the jeep rocked to a stop, Paul stammered, "Ah...well, Hamilton will explain that. Let's go inside; then, you can talk to him."

When the two brothers entered the conference room, Hamilton and another man rose. The tall, thin stranger wore a tailored western-style suit. His distinguished gray-tinged hair, coupled with mirrored sunglasses, drew Nolen's attention.

"Glad to see you again," Hamilton greeted the two brothers. "Allow me to introduce our newest partner. Senator Santreno, this is Nolen Martin, the head of our Association."

Nolen's eyes shifted back and forth from the banker to his brother, signaling his irritation. Then, he shook the stranger's hand while Hamilton continued with the introduction.

"Senator Santreno chairs the committee which oversees the Nevada Gambling Board. He learned of our licensing efforts, and has suggested several ways in which he might be of some service. Since he has many influential friends throughout the state, we offered him a partnership." Hamilton nervously smiled at Nolen. "We were unable to contact you in San Francisco when this opportunity arose, so we went..."

"The same percentage offered to everyone else?" Nolen interrupted.

"Yes, yes, that's right," Hamilton stuttered, as he bobbed his head up and down.

"Senator," Nolen quizzed, "do you understand what is taking place here?"

"Of course. Building a boomtown—before the boom—is an ingenious venture from which we can all profit. However, Mr. Martin, my vital support is worth far more than eight percent of the mine."

Nolen's internal anger escalated. He cleared his throat. "Ownership in the mine is not unlimited, Senator. You received the last uncommitted partnership."

"Someone else should give up a portion of their share. After all, it would be a shame if you lost your casino license, or if the Nevada police began halting your delivery trucks. Even unforeseen labor problems could delay your construction schedule."

Nolen's jaw muscles tightened. He squinted at the audacious politician. "Let's cut this song and dance crap, Santreno. I formed this Association with partners contributing specific skills that enable them to fulfill critical tasks. Tasks which will capture different portions of the vast fortune available to us. Each person must willingly do their part; otherwise, we'll all lose."

"Your ideas appeal to me, except the percentage split."

"Obviously, you can help us, so you should reap a fair share of the benefits. And each of the other partners will gladly contribute to your re-election fund. But we're over halfway toward our goal. Frankly, I'm out of time, maneuver space, and contingencies. I can't afford partners who cause problems that endanger our success. So, let me be specific—in terms you can clearly comprehend." Nolen looked down at the table, as he flicked a speck of dust off the vinyl top. "You can be very rich, or you can be very dead."

Santreno haughtily responded. "That's pretty tough talk, threatening the life of a state Senator."

"You mean the life of a greedy, corrupt state Senator. You understand power and have toured this town. You've seen the men working here." Nolen jabbed a thumb over his shoulder in the direction of the center of the compound. "Do you think I could find ten men who would willingly kill you, if I offered them $100,000? How many would jump at the chance to slit your throat for $200,000, then fly to Australia, where they would receive $50,000 for each of the next twenty years?"

"I'm not intimidated by threats."

"Don't let your title cloud your judgment. These aren't the slick halls of the state capital. The game is played much differently here. We're using the rough rules that many of your Italian ancestors grew up fostering. Knock off the bullshit! Forget about upsetting the partnership. That isn't wise for you, or for the Association. Now, what's your decision."

The Senator removed his sunglasses, coldly staring at Nolen. While cleaning the lenses with his handkerchief, Santreno finally smiled. "Well, Paul, you were right. Your brother is a rather convincing fellow. Nolen, I can see you've thought this through. You are correct. There's no need to butt heads, when I can make money just by ensuring this project is not interrupted."

As the group relaxed, Hamilton glanced at his watch. "Nolen, I urged the Senator to come here, because I felt you should meet the man who will provide us with political support inside Nevada. However, we must avoid people connecting him with the town. He is due back in Reno in an hour, so it is best if he departs soon."

"We can schedule another meeting later this week," the Senator interjected.

"I understand," Nolen said. "Hamilton, make the arrangements."

The banker and the politician left the room as Nolen turned to his shorter brother. "Paul, the next time a situation like this arises, could you and Hamilton delay committing us, until I'm informed?"

"You hired us to do a job, which sometimes entails making hard decisions. Especially, when you're not available. Hell, we had no inkling he was going to throw us a curve. He never mentioned anything about wanting a larger share than the other partners."

Nolen shook his head, frustrated with Paul's naiveté. "Brother, you have to expect the worst from people, when a great sum of money is involved. This twist complicates things, brings in a stranger who may endanger our efforts. Even worse—our reserve Association shares were consumed in this bargain."

"It also provides us some powerful benefits," Paul quickly reminded Nolen.

"True, but I don't trust him. He's the sleazy type of fat cat that I hate dealing with. Everyone else in the Association comes from a background I can relate to. They are people who have struggled to make a living, people motivated to see the Association succeed."

"Your talk of murder scared me," Paul complained.

"I was bluffing, trying to get that sleaze to come around."

"It didn't sound like a bluff."

"Then maybe it will work."

Paul wiped sweat from his brow. "And if it doesn't?"

"Shit, I don't know. I'll figure something out later. Come on, show me the town."

They returned to the jeep, then drove to the center of the compound. There, Nolen began entering every completed and partially finished building. He asked each foreman if needed materials were arriving on time. He quizzed workers, probing for any recommendations. While walking through the southern portion of town, he directed warming tents be erected, so the evening crew could periodically receive relief from the wind, to enjoy a cup of hot soup or coffee.

Two hours passed. The brothers concluded their inspection with Nolen being introduced to a team of five men. They had installed the town's perimeter fence, five days ahead of schedule. Nolen made a production of counting out a $3,000 bonus to the group, for their exceptional accomplishment.

While the two brothers walked back to the jeep, Paul searched for Nolen's approval. "Well, how about some feedback, impressions, opinions?"

"The town is progressing very well," Nolen commented. "I'd say you have this project under control. Great job so far. But, will you finish on time?"

"It's going to be tight. However, the crews have been picking up the pace, as they have learned the work routine. Plus, they are responding well to the bonuses. We appear to be rolling now and should meet our deadline."

Boodan stood in the ebony shadow cast by a tall stack of lumber. He peered through a side window, viewing the old caretaker seated in the entrance room to the workers' sleeping quarters. Impatiently, Boodan watched the slight man's head bob up and down. The caretaker weakly fought against falling asleep at the end of a long day, as he waited for his replacement to arrive.

When the old man's chin settled onto his chest, Boodan lifted a five-gallon can of gasoline, wrapped in an empty cement bag, and strode to the entrance door. He slipped inside and tiptoed up the stairs leading to the darkened second floor. Downstairs, more than 300 exhausted men slept. But the upper floor was empty, since the night crew was undergoing Nolen's inspection in the southern part of town.

Boodan quietly pulled blankets and mattresses off several beds. He strategically laid them along the center aisle, then doused the material with a generous slosh of gasoline. Hurriedly, he used a knife to cut several sheets into strips. He trailed a four inch wide line of white cloth from the soaked bedding to the doorway, located at the top of the stairs. He splashed gas along the strips of sheet. This wick, he thought, had better be long enough to allow me time to flee the building before the fire reaches that half-filled can of gasoline. When the gas can ignites, there's going to be one hell of an explosion. That will catch Martin's attention.

Before Boodan could light the edge of the wet wick, footsteps sounded on the stairs. Jerking upright, he back-stepped, hiding behind a bunk. He waited. Seconds later, the caretaker limped through the doorway, illuminated from behind by a single bulb dangling at the top of the stairway. The old man sniffed the air, scanning the dark room with his flashlight. When he spotted the jumble of mattresses, he forged deeper into the sleeping area. "What the hell's going on here?" he muttered.

Boodan smashed his fist into the side of the caretaker's jaw, knocking him against a wall locker. The old man crumpled behind several foot lockers at the base of the bunks.

Boodan swore under his breath, hoping the sudden clamor had not alerted anyone sleeping downstairs. Hearing no other movement, he lit the long wick, watching the yellow flames lick along the white strip. He then scurried down the stairs and into the welcoming darkness. The big man ran several hundred yards to the safety of his pickup, not looking back, nor worrying about the innocent men sleeping below the gas bomb.

At the truck, he took a pair of binoculars from inside the cab to scan the compound, searching for Nolen's jeep. He spotted the bouncing vehicle, returning from the southern part of the town. Now it's my turn, Boodan gloated. In seconds, Nolen, you'll begin paying for going against my cousin and me. And if you get separated from the guards, you're mine!

The staggering explosion blew out all the second floor windows, and blasted a red and yellow fireball through the tin roof. Dancing light illuminated the entire town. "Oh no!" Paul yelled.

Nolen gazed at the unexpected destruction, momentarily stunned by the flashing fury. Then, he jumped into action. "Paul, I'll lead the security team into the building. We'll get everyone outside. Go! Get bulldozers and water trucks! Have the water trucks wet down the store and warehouse. We can't save the sleeping quarters, but we must salvage the supplies stored in the other buildings. Do you understand?"

"Oh my God, how many men will be killed? This is terrible!"

"Paul, get moving—now!"

"How could this happen? We'll be ruined!" Paul was slipping into hysteria.

Nolen slugged his brother in the chest. "Get those bulldozers, damn it!"

Paul stared at his brother.

"Use the dozers to knock down the walls of the sleeping quarters. Push the debris away from the store and the bank."

"All right, I'll get it done."

Nolen ordered one security guard to organize the workers who were escaping from the inferno. They were to first assist anyone who was injured, then begin fighting the fire. Nolen signaled to the rest of the security team to follow him into the burning building. Two did, while the others faltered.

Billowing, choking smoke spewed from the second floor, as streams of men gushed through the first floor doors and windows. Most men were alone. Many were dazed. Some carried burned co-workers. A few, full of terror, knocked down weaker men in their desperate scramble through the exits.

Once inside, Nolen sent the security team to search the first floor area to ensure everyone had escaped. Meanwhile, he ran up the charred stairway, hoping to verify no one was upstairs. Searing heat funneled through the splintered doorway, as Nolen reached the top landing. He could barely face the furnace of fire consuming the sheet rock and many mattresses. As he turned in retreat, he spied the old caretaker, bloody and coughing. The old man was sprawled behind several foot lockers, which had deflected the explosion from flash burning his body. Nolen leaped to the man, snatched him into his arms, and charged back down the stairs. Nolen ignored the blisters rising on his shoulders and neck.

Outside, several men hastened to Nolen's aid. They lifted the caretaker from him. Nolen fell to his knees, choking and gasping for clean air. The security team leader rushed to Nolen. He explained the blast had only ruptured the ceiling of the first floor in one area. The majority of men sleeping downstairs had been protected from the upward eruption of the explosion. It appeared none of the day crew had perished in the blast. Nolen felt somewhat relieved.

He struggled to his feet and ordered a bulldozer to smash the building. The tractor rumbled forward, then slammed into the corner of the flaming structure, shoving the wall support over to a forty-five degree angle. The bulldozer reversed, then repeated its assault. Water trucks screeched to a halt. Quickly, they began spraying the buildings nearest to the fire. Clouds of steam gushed from the nearby store and bank when the streams of water cascaded over the hot metal.

Paul's jeep skidded to an abrupt stop next to Nolen. Paul hollered, "Are you all right?"

"Yeah, just some burns. I'll be OK. No one was killed. We were lucky, very lucky."

The sleeping quarters crumpled with a roar. Paul shook his head in disbelief, as he watched millions of wild sparks swirl into the air. "It had to be arson! Nothing inside the buildings is volatile enough to cause an explosion. Someone attacked us!"

"You're right, Paul. I carried the caretaker from the top floor. When you question him, I expect he'll verify our suspicions. I bet CMR is behind this!"

"But, Nolen, my people have been checking who we hire. Maybe you underestimated Santreno's nerve?"

"I'm not blaming you, Paul. With the number of employees we've recently put to work, something like this was bound to happen. At least, now we know enemies have infiltrated the compound. We must prevent another strike! Immediately, check every building for any other arson bombs. Schedule guards to patrol each building and the storage yard, twenty-four hours a day. Get those guards in place within fifteen minutes."

"OK, Nolen. Later tonight, I'll visit any men we transport to the Susanville hospital. Tomorrow, I'll hire a force of doctors and nurses, for on-site medical assistance."

"Now you're thinking clearer, brother," Nolen encouraged Paul.

"We'll also need to provide another sleeping area for our men. This loss will surely delay our schedule! That's probably what the saboteur intended. Our plan is not going to work now!"

"Calm down, Paul. We can handle this. Since both flop houses are nearly ready, we can shelter the men there, until you rebuild the sleeping quarters. Put construction priority on replacing this building."

Paul hopped from the jeep. He held the door open. "It's past time for you to get away from here! If CMR does have an infiltrator in this compound, they may try to kidnap you."

Nolen hugged his brother. "Be careful. Remember, you're more important to me than this town."

Boodan lowered his binoculars, after watching the armed security team shepherd Nolen to his waiting airplane, and ensure it safely departed.

He clicked on the CB radio transmitter, hanging from the bottom of his pickup dashboard. "Luke, talk to me. Boodan's got his ears on."

The reply crackled back. "This is Luke. What's on your mind?"

"Relay this message to Tower. The main man was here. Flew in, looked the place over, then took off."

"You sure it was the guy we're looking for?"

"Positive."

"How vulnerable is he?"

"Gonna' be tough to get close to him. Guards were around all the time he was here. To grab him, we first gotta' separate him from his security. Then, we could use a helicopter to snatch him before someone stopped us. Or, he could be dropped real easy with one rifle shot."

"OK, I'll pass the info to Tower."

"Did you find the yard that bought Jorgenson's pickup?"

"We're driving there now."

"Give me directions, and I'll meet you."

Boodan's crew of former felons scurried about, disrupting the usual nighttime loneliness of the salvage yard. Boodan watched, as the stack of debris was finally illuminated by auto headlights. Digger's wrecked truck perched atop a pile of car bodies and chunks of other rusted equipment. The Cajun wondered if Hilda Jorgenson's suggestion to search for her husband's prospecting notebook would pay off.

One man hauled a chain from a wrecker, then attached it to the frame of the crumpled pickup. The truck creaked and rattled, as it was jerked from the surrounding metal mess, and dragged onto an open spot of greasy oil slicks and brown mud. Two of Boodan's men attacked the bottom of the truck. Within seconds, one of them discovered the hinged plate. He ripped it open with a crowbar. The old prospector's notebook plopped into the mud. Eagerly, Boodan snatched up his prize.

30

Takeover

24 September

Trimmed hedges and a lawn resembling a putting green added a richness to the manicured hillside residence overlooking Riverside, California. A cool morning breeze, blowing inland from the Pacific, kept the usual smog out of the valley below. It shook the tall palm trees bordering the two-story, Spanish style mansion owned by the president of Benjamin Mining Company.

Maida parked the car across the street and placed her hand on Nolen's arm. He sat, with his eyes closed, attempting to ignore the dull, steady pain pulsing from the half-healed blisters tattooed across his neck and shoulders. A sourness rumbled within his stomach. He concentrated upon the song of a bird, happily chirping in the nearby shrubs. Slowly, his efforts prevailed, and he began to relax.

"Well," Maida said, "we've spent most of our money, maxed our credit limit, and now have only four days to complete everything."

Determination creased Nolen's tanned face. "Yes, it's time we start playing the end game, learn what the rest of our lives will entail. Will I see you happy, draped in diamonds and furs?"

"Wonderful idea. I like it."

"Or crying and broke?"

"Stay positive, Lover."

Nolen smiled, despite his concern. "OK, let's weave a convincing story to net this fish."

They approached the expansive oak entrance door, where they waited for someone to answer their ring. A uniformed Mexican maid opened the door. "What may I do for you, Señor y Señora?"

Nolen answered. "We've come to see Roger Benjamin, please."

"Señor Benjamin is not expecting any visitors this Friday. Please call his office Monday to schedule an appointment." She attempted to close the door.

Nolen quickly handed the maid a sealed box. "This is my calling card. Please give it to him. He will recognize its importance and want to speak with us."

The maid looked into the faces of the strangers standing on the doorstep and gently shook the small container. Its weight pressed heavily into her hand. "Humph," she grunted, then shut the door.

Maida glanced at Nolen and shrugged. Minutes dragged until the woman returned. "Follow me, por favor."

As they walked down a hallway, the trio crossed luxurious Persian rugs and passed life-sized marble statues. The design of the building became apparent. It was U-shaped, with two spacious wings projecting from the central structure, forming a wide courtyard which they entered. Tall bougainvillea bushes entwined the wooden columns supporting the second floor balcony. Their thick vines draped brilliant red blooms above the entryway. Dwarf citrus trees were spaced along the walls. And large redwood planter boxes, containing hundreds of multicolored flowers, scented the patio.

At a glass table, sat a weighty executive. His face was accented by his bulbous nose, and skin scarred by acne. He was fingering the contents of the small paper box, while sipping a cup of coffee. "Two pounds of gold ore, wrapped in an assay report, is an unusual calling card. You Martin?"

"I am."

He motioned toward several cushioned chairs. "Have a seat."

The mining executive nodded to Maida. "Your secretary's a cute little honey."

"I'm his partner and stock broker," Maida corrected him. "I'm not his cute honey." She placed her attaché case on the breakfast table.

The mining executive raised his eyebrows. "Real sharp tongue for a lady."

"You're acting like your chauvinistic reputation," Maida answered. "Let's overlook that, and get down to business."

"Sure. Martin, you've tweaked my interest. Tell me your story."

Nolen began. "I found a vein of gold, one and one-half feet thick, ten feet wide, and at least three-quarters of a mile long. As the assay report specifies, it is eighty-five percent pure. We need a mining company to extract the gold. You're my first choice."

The man's nostrils flared, as he burst out laughing. He pounded his chubby hand against his chair. "Damn, that's the best story I've heard in some time. One thing about prospectors and geologists, they are always positive. The future is rosy. There are no worries. We're certain to strike a rich deposit."

"This time, it's true. I can prove it," Nolen insisted.

"Sure. The only problem is your incredible imagination. You exaggerate beyond all normal bounds. So, pardon me for not taking you seriously, when you claim to have found the largest gold deposit uncovered during the last 150 years."

"I'm as serious as a heart attack. That sample you're holding came from the vein. One of your own people can verify
the assay."

"To judge any offer," Benjamin replied, "I must first hear
all the business conditions, assuring me a profit. I require
new partners to commit enough of their own assets, so they
share the risks and are motivated to succeed. As a policy, I
demand significant control of all operations. If those conditions are met, then and only then, do I begin to feel my company isn't being conned."

"Public records," Nolen replied, "indicate you and two
board members control forty-six percent of the voting stock
of your company. The split is twenty-six percent held by
you, and ten percent owned by each of the others. I'll trade
you six percent of the mine in exchange for fifteen percent of
Benjamin Mining Company preferred stock. My six percent
offer allows each of your three major stock owners to acquire
equal shares in my gold mine."

Benjamin unclasped his large hands to clean a fingernail.
"If this story is true, or even partially true, six percent of the
mine is worth a fortune. The company stock you are asking
for will rise in value seven, maybe, ten times. Compared to a
portion of the vein, that would be peanuts. However, my
two partners and I would drop from owning forty-six percent of the voting stock, to only thirty-one percent. Making it
still feasible, but more difficult, to control the company."

Nolen persisted. "The stock swap also compensates your
firm to mine the site and process the gold for the next year.
Also, you must buy from our Association, fifty acres of land
on which you can build a smelter. That acreage is located
within twenty-five miles of the strike. It will cost you an additional $200,000 cash. Monday, you supply the men and
equipment to secure the gold site. Until we file, you maintain strict confidentiality concerning our activity."

Benjamin sipped his coffee, contemplating the offer. "Sounds like you've considered some of the realities of opening a new mine. However, I'm still skeptical. Over the years, too many people have tried to fleece my wallet. And you seem like you could con a virgin out of her virginity. Why my company?"

"Your firm," Nolen answered, "has the resources and enough experience to do the work. In addition, I'm sure you'd enjoy beating Continental Mining and Refining to this strike. I believe they're your prime competition."

"Competition my ass! They always push the rules of the game beyond the edge."

"Just like you do." Nolen smiled. "So, you should be tough enough to back me during this effort."

"Do they know about the discovery?"

"I began hunting for this vein six weeks ago, based on two small discoveries made during the previous year. CMR started trailing me, watching to see if I would hit a larger strike. They learned from an assayist that I had found a rich vein. Several of their teams are undoubtedly scouring the region where I made the find. If we don't act quickly, they might hit it."

"Very well," Benjamin replied, "I'll continue to listen, but I need guarantees. I want some standard business practices to protect me from fraud. I'm not committing a cent, until I have proof—damn substantial proof!"

"Of course," Maida calmly concurred, while removing three sheets of paper from her attaché case. She handed one to the executive. "Here's our single page contract containing our stipulations. We can add additional safeguards as necessary. We will keep it brief and specific, using clear language. That way, it will be legal, and there will be no confusion.

"We anticipated your need to personally inspect the site, electronically survey the vein, and then assay ore samples. Please, notice the phrase which requires you to complete the stock transfer within twelve hours after extracting core samples. That will provide you with sufficient time to verify the value of the strike, prior to obligating the assets of your company."

"If I decide to commit, you can bet my men will take hundreds of samples. I've seen salted mines before."

"We encourage you to do just that," Maida affirmed. "For your added security, we inserted paragraph III. All Benjamin stock will revert to you and to your partners' control, if the Association does not fulfill the following conditions. The first delivery of ore must be completed within fourteen days after you transfer stock to the Association. And the first delivery must contain 1,000 troy ounces of gold."

The executive read the document. The silence was interrupted as he cracked his knuckles, while he deliberated the offer. "You must know, a good gold mine only provides one ounce of gold for every ton of rock extracted. Which means, you're either full of bullshit, or you really do believe your vein is incredible. I can't imagine anyone delivering such valuable ore."

"We are confident in how good our vein is," Maida responded.

"I demand additional protection. Add a clause specifying that the stock transfer occurs only after I also verify that a legal filing on the site has been recorded. Then, and only then, will I transfer a portion of my company to you. Otherwise, I could be left with nothing of value, if you can't secure a valid claim."

"There is no problem to write in that addition," Maida stated.

"Furthermore, I will sign the contract only if it is edited in another way." Benjamin tapped his chest with his thumb. "Stipulate that you are selling the six percent ownership in the mine to me—not to anyone else within my company. I don't need to share such wealth with my minor partners. Those twits wouldn't know what to do with it anyhow."

"Sure," Nolen replied in his most agreeable manner. "Now, this second sheet lists the men, material, and equipment we need available immediately. The third sheet details the work and time schedule. Have your people at the Orange county airport on Sunday. We'll fly to a pre-staging location. On Monday, you'll accompany me when we file, travel to the site, and extract core samples."

The three negotiated for thirty minutes more, finalizing their plans. Then, at Nolen's insistence, two security men arrived to guard the newest Gold Association partner. Nolen and Maida departed. They drove across Riverside to a much less pretentious residence owned by Benjamin's senior field engineer.

Inside the man's sunny, ranch-style home, Nolen spoke with the Production Chief. "Mr. Kemper, would you like to be President of the Benjamin Mining Company? I need someone who is competent, someone totally trustworthy, and someone willing to cooperate with us. I believe you're that man."

Karl Kemper frowned and shook his bearded head. "When your lawyer telephoned to arrange this interview, I gained the impression you would be making a serious offer for me to manage a mining operation. Be assured, there's no way in Hell that Roger Benjamin would ever relinquish control of his company to me."

"Why?" Nolen asked.

"He hates my guts. You see, I helped his father set up the company, then guide its growth. Benjamin senior was a super guy, but his kid is a spoiled brat. Roger worked for me one summer when he was in college. He got his ass beat by another crewman. The kid instigated the trouble and deserved the results. He came to me, demanding I fire the other guy. I didn't agree. Instead, I told him I would throw him off the job if he didn't stop causing trouble. We haven't had a civil relationship since."

"We are aware that his arrogant ego gets in the way of many of his relationships," Maida agreed.

"Luckily for me, when his dad died, I inherited five percent of the company stock. Benjamin senior stipulated in his will that the company must retain me as Senior Production Manager, until I chose to resign. The old man knew what he was doing. I'm the one who keeps the business operating in the black. Just this year, I've kept Roger from screwing up on three separate occasions."

"We've heard you've been doing that for years," Nolen added.

"Makes no difference. Roger and the other two board members control the company. Anything they want, goes. And Roger remains his own fat, obnoxious dictator. I can't name three people who can tolerate him for more than an hour. So, unless you plan to buy the company and terminate Roger, I'll never be President."

"Accurate conclusion," Nolen responded. "Forty-five minutes ago, Roger Benjamin sold our Association fifteen percent of the stock in his company. What he doesn't know is that during the last several weeks, Miss Collins has gradually purchased twenty percent of the outstanding stock on the open market."

"Why."

"Adding your five percent, we immediately control forty percent of the voting stock. With that much stock, and by influencing the other board members, we'll eventually control over sixty percent of the voting rights. The company is ours, but Roger doesn't know it! Does my offer sound serious now?"

The Production Manager appeared surprised. He ran his hand slowly across his beard. "Yes, it's beginning to. But I need to see some proof that you own the amount of stock you claim. Furthermore, explain why you want the company. And how do you plan to run it?"

Nolen reached into his coat. He removed the contract with Benjamin's signature. From another pocket, he produced a hand held tape recorder. Nolen clicked the recorder on and watched as Kemper listened to the recent negotiations. Then, Maida withdrew several bundles of stock certificates from her leather briefcase.

When Nolen and Maida left, Karl Kemper had become a five percent Association member. "Let me see," Maida said, as she drove toward the Riverside airport. "That's the ninth and final partner, unless you have some more surprises up your sleeve."

"He was the last one. We've brought on board every key person I originally planned for, and some that I didn't."

"Well then, that accounts for 100 percent of the mine shares. Six partners each own eight percent. Benjamin has six, and Kemper owns five percent. Leaving you, only forty-one percent."

"Unfortunately you're right," Nolen replied. "When I was forced to include Senator Santreno, I lost outright control of the mine. However, only you and I know that fact."

"In the future, someone might challenge how you run the Association. You could be out voted, unless you form a coalition with at least two other partners. Have you considered that?" Maida questioned, voicing her concern.

"I'm hoping no one will question my leadership just yet. We have established the companies needed to support our plan. Also, the boomtown is almost prepared to receive the gold rush. So, before the other partners learn that I can be overruled, there's only one major event remaining. On Monday, at 3:30 in the afternoon, I'll file our strike in Susanville."

"I'm pleased you took my advice," Maida replied. "The New York Stock Exchange will be closed by then, giving us a night to fuel the market with news of the massive discovery. Personally, I'm weary of waiting. These last few weeks have been nerve-wracking."

"I realize, pretty lady, that the Association tasks have been demanding. But the difficult work is completed. This afternoon, all you need to do is notify the other partners of the filing. Then, I want you to enjoy a relaxing weekend. Don't go back to your condo. I've reserved a room for you at the Orinda Spa with the best court times scheduled at your favorite country club. The bill is paid. Go pamper yourself. Rest, and wait for the fun to begin, Monday afternoon. Next week, you'll be a millionaire."

"Nolen you're so considerate. Come with me. We'll pamper each other."

Nolen took Maida's delicate hand in his. "I can't. I must get into position near Susanville. After the filing, we'll spend as much time together as we want. No more stolen evenings or hurried embraces. Maybe...maybe we could live together?"

Maida's shock caused her to veer into the on-coming traffic lane. She quickly recovered and wheeled the car to the curb. She slipped into Nolen's arms, thrilled with the anticipation of his nervous request. "You've been a perpetual surprise to me, ever since I first saw you flopping out of the pool. You're continually attempting to control your emotions—and continually failing."

"I think I've been doing just fine," Nolen protested.

"Everyone else isn't as close to you as I am."

"So, what is your answer?"

"I'll accept living together for now. But I expect more commitment than that from you."

Nolen grinned. "I suspect you do. Just trust me, pretty lady. Soon, nothing will keep us apart."

31

Timing

Pam Pendleton crumpled the sheet of writing paper she had been working on, and threw it into the nearby wastebasket.

Her bodyguard looked up from across her plush office. "That's the eighth time since lunch that you've tried that piece. You're going to kill a tree at the rate you're going."

"Yeah, well, it's none of your business." She resumed rubbing the dull, throbbing pain in her hand that Jake Sampson had broken. Each time she recalled that terrible night, her anger flared at Nolen, and at his security henchman. But she knew well that that incident was not causing her nagging, bitter mood.

I may be out of sorts, but I'm certainly not pregnant, she thought. Ever since my second abortion in college, and the subsequent infection and uterus scarring, I've been sterile...couldn't bear children, even if I tried.

Ironically, that furthered my career. I would never have received the better assignments, had I rushed home every night to care for some snot-nosed kid. When I joined this industry, there was precious little acceptance of working mothers.

Anyway, my sour mood isn't due to the lack of a sex life since joining the Association. And chances are, I won't have any until I'm relieved of my set of twenty-four-hour spies, hovering over me since Nolen caught me stealing that trunk of gold. Hell, that doesn't matter. I wouldn't have had any sex, even without the guards. At least, not anything satisfying. After all, I don't have a decent, affectionate relationship with any man.

This recurring depression is the consequence of a missed period. The second time in four months! I'm beginning menopause. Why me? I'm young! Or at least, I used to think so. Here I am, at age 47, getting older by the hour, with no caring husband, no close family, or even dear friends. Just my token position at this damned TV station. The position I was demoted into a few short months ago.

"Damn. What a swell obituary!" she exclaimed.

Her bodyguard stared at her again.

Pam flipped him the bird, then jerked a cigarette from her purse. She heard him faintly mutter something about a "no class witch." The telephone rang at her desk. "Pendleton here," she barked into the receiver.

"Pam, this is Maida. I have some important news for you."

"Well, it's so delightful to hear from you, little Miss Muffet. What have you been doing?"

"This morning, Nolen and I finished our work in Los Angeles, and he selected the filing day and time."

"Did you two satiate yourselves while you were together?"

"If you mean, did we sleep together, that's none of your business. He did, however, give me a nice gift."

"Oh, he bought you a vibrator?"

"Just a weekend stay at my favorite spa, with tennis lessons at the country club."

"What girls' camp is this?"

"The Orinda Spa. Pam, why do you always pry for information?"

"You don't have to act so miffed. As a reporter, it's instinctive for me to be inquisitive."

"I'm bored with your innuendoes. Nolen is expecting you and your crew to be in Susanville, no later than noon Monday. The filing is set for 3:30 p.m. Make a note."

"Don't worry, I remember the plan. Rest assured, I'll take good care of Nolen after the filing. After all, you'll be busy in the bay area, won't you? See you in a week or two, Sweetie." Pam dropped the phone into its cradle, enjoying her small triumph.

Concern fueled Pam's negative mood as she pondered Nolen's risky scheme. If Nolen is successful in securing control of the mining site, then the Association partners will become wealthy. However, CMR did not become a mining powerhouse through sheer luck. That conglomerate is a formidable opponent, with extensive experience in the legal aspects of the exploration business. CMR could easily destroy the Association in the courtroom.

Pam puffed her cigarette, as she continued to analyze her position. I would feel far more comfortable establishing some arrangement with CMR, just in case they should win this contest Now, all I have to fall back on is the exclusive news story of the gold discovery. That story should return me to star ranking among TV news anchors. Nonetheless, something could still go wrong prior to the filing.

Pam paced back and forth. Each time she reached the windows, she would pause, tap her foot, adjust her clothing, then resume walking and plotting. Two chances to succeed out of three are good, she concluded. Three out of three would be best. That way, no matter what the outcome, I'll be a winner.

Abruptly, she spoke to her guard. "Excuse me. I'm going to the lady's room." She stepped through the door of her office and signaled for her secretary to follow her, before the guard sauntered from the room behind her.

Both women walked down the carpeted hallway into the restroom. Once inside, Pam scribbled a note and gave explicit instructions to her assistant; then, Pam returned to her office. The guard dutifully trailed after her.

Ninety minutes later, Pam's intercom light blinked. She pushed the flashing square and inquired, "What is it?"

"Miss Pendleton," her secretary announced, "the station manager wants to speak with you in his office. It concerns the TV crews."

"I'll be right there." Pam glanced at her guard. "Sorry, chum, but it would be difficult explaining you to my boss. You'll have to park outside his office until I'm through." Purse and notepad in hand, Pam walked the few paces past her secretary's desk to her supervisor's corner suite, with its panoramic view of the bay. She quietly locked the door behind her, relishing the fact that she had outsmarted her bodyguard.

The station manager had taken the afternoon off to play golf, as was his practice every Friday. Pam often took advantage of the comforts of his office when he was away. The plush atmosphere of tufted leather and expensive art were congruent with her taste for luxury. Now, Tower and Boodan stood waiting within the spacious office. Pam strode past them, then wiggled into her supervisor's chair behind an oversized basalt black desk.

"Mr. Tower, I'm Pam Pendleton. I sent you the message. Incidentally, you look exactly like the photo your company provided to our Business News Division, three weeks ago. Sometimes file pictures are..."

"So, when is Nolen Martin filing?" Tower demanded.

"Obviously, you did not believe my note. I do not intend to supply critical information regarding Nolen's activities free of charge."

Boodan's face turned ugly, as he stepped toward Pam. "Don't get coy, bitch! We could easily take you outta' here, and force you to tell us what we want to know."

Pam slipped her .25 caliber pistol from her blazer pocket and cocked the hammer. "Wise of me to bring my little equalizer, don't you agree?" She first glanced at Tower, then returned her steady gaze to Boodan. "Mr. Tower, let's not begin our friendship on the wrong foot. I've been with Nolen since he brought in the first load of gold, which your people failed to steal. Now, I'm offering to become your most reliable source of information. But your muscle-bound helper is acting crude."

Tower pointed toward a leather chair stationed near the main door. "Sit down, Boodan, and be quiet. Miss Pendleton and I need to come to an agreement."

Boodan sullenly retreated to the chair, all the while considering ways of relieving the petite woman of her gun.

"Fifty thousand dollars for today's information," Tower said, "and $10,000 each time you pass something to me in the future, is a considerable amount of money."

"Not really, since my information will eventually allow you to control Martin's gold discovery." Pam removed a one-page contract from her purse, then slid it across the desk for Tower to review. "Furthermore, when CMR files on the claim, this publicity consulting contract will go into effect. CMR will begin paying me an annual fee of $100,000 for five consecutive years. Here is my pen. Won't you have a seat?"

Tower sat across from Pam and tugged a fat envelope from his suit coat. He removed the fifty, $1,000 bills in the envelope and spilled them onto the desk. "You better not be lying, or working for Martin, or trying to fool me. If you are, I'll send Boodan back for a visit. He might, let's say, re-negotiate our agreement, some lonely night when you walk to your car. You wouldn't want that to occur, would you?"

"Sign the contract," Pam coldly responded, not frightened by his threat. "Our arrangement must appear totally legitimate. Especially, if you later instruct your corporate lawyers to not honor it, forcing me to sue you."

The mining executive scratched his signature onto the document. He tossed the sheet back to Pam. "Prove you're worth the money I just paid you."

"Certainly. Nolen will file his claim on Monday, three days from now, in Susanville, at 3:30 p.m."

"If Martin does not attempt filing Monday," Tower warned, "Boodan will visit you that night. Where can we locate your partner?"

"I don't know. He moves around constantly, a maneuver designed to stay away from your people." Pam smiled as a delicious thought overcame her. "However, there is someone... Maida Collins, his girlfriend, works as Nolen's information coordinator. I'm sure if your goon here were to prod her, she would gladly cooperate. She will be lodging at the Orinda Spa this weekend."

"That's helpful information, since Miss Collins keeps eluding us." Tower turned to glare at Boodan. "She surprised my men the other day by wheeling into the San Francisco airport. She boarded a private plane, just prior to take off. Tower returned his attention to Pam. "Keep talking, you're beginning to earn your money."

Minutes later, Pam left her supervisor's office, bound for the company cafeteria, with her guard following at her heels. Pam's secretary then ushered Tower and Boodan through a side hallway exit.

As his limousine blended into traffic, Tower rubbed his hands together. "Boodan, my luck still holds. The unexpected information which this Pendleton broad has provided fits with what we know concerning Martin's activities."

"Yeah, I think she told the truth about the filin' date."

"Finally, we have the initiative. We know the what, when, and where of Martin's next move. All our planning will pay off this coming Monday. I cannot wait to see the look on Nolen's face. Can you have everything prepared by then?"

"No problem," replied the Cajun. "Let me use your car phone. I'll call my men in Reno and get them movin' right away. The paperwork is already prepared. I'll warn the County Recorder to expect the filin'. We'll check that our microphone system works at her office. And I'll have old lady Jorgenson nearby on Monday. Also, Pendleton's suggestion to grab that stock broker is a good one."

"Why now?"

"Martin is committed at this point. The town he's been building is nearly done. All his eggs are about to hatch. So, timing is very important to him. Snatchin' Collins this weekend should throw him off balance. It'll give us an additional bargaining chip before Monday's filing."

"H'm, yes, that is a good idea. It's about time we applied some direct pressure on that shithead! I have been dreaming of using the broker to punish Martin. We could tie her up, and force her to tell us everything she knows about Martin, his mine, and his desert town." Tower licked his lips in anticipation. "Boodan, how familiar are you with Collins' routine?"

"She's an athletic girl. At least once a week, she does her girlie aerobics, or tennis, or golf. When I was figurin' ways to kidnap her, I got CMR to buy an Orinda Club membership. That way, we could stay close, even when she's exercising at her fancy playground."

Tower again rubbed his hands together. "So, if Collins returns to San Francisco, as Pam indicated, she'll be vulnerable. We should be in position tomorrow, when she plays tennis."

"Right."

"But doesn't she still have a couple of bodyguards protecting her? You sure it's only two?"

"Yup," Boodan confirmed. "They always check the people around her, every time she's out and about. We've had to keep our distance, so we wouldn't be noticed. But we're well prepared to handle them."

"Have you tested the explosive devices?"

"Yeah, they'll easily blow their tires. Also the car we bought looks just like the one her bodyguards drive. She won't suspect a thing."

Tower stared through the windshield, oblivious to the traffic nearby. His silence lasted several minutes. "OK. Tomorrow morning, implement your plan."

Boodan smiled a smile of delight. "Right, cousin. I'll be in touch, after I get my hands on the little bitch."

32

Surprise

25 September

Jake Sampson rushed down the narrow central hallway of the boomtown headquarters, and barged into Nolen's office.

Nolen looked up from the numerous reports and schedules on his desk. "Jake, I know you're a big guy. But try leaving the doors on their hinges."

"We have a serious development, requiring an immediate decision!"

"I don't need any more ugly surprises, Jake." Nolen exaggerated, attempting to remain calm. "Tell me I'm lucky. Something simple has gone wrong, like the town's water supply is poisoned. Or all the workers quit."

"Maida has been kidnapped."

"Oh, no!" A wave of nausea washed through Nolen's body. "How do you know she's been kidnapped?"

"I just received a call from her guards. About forty-five minutes ago, while they were following Maida's car, a man in a blue van used a transmitter to set off explosive charges attached to the tires of the bodyguards' car. That's when our men lost contact with her. Since then, they found Maida's sports car abandoned on a sparsely populated section of road leading to the spa. The keys were in it, and the left front fender was smashed. They probably forced her off the road, then grabbed her."

Nolen raised his fist and slammed it onto the desk top. "Damn, damn, damn. It's got to be CMR! I knew all along Maida was our most vulnerable partner. Damn, I thought I had protected her." Nolen closed his eyes and shook his head, worrying. "Any blood trails?"

"None."

"No one, except me, should have known where she was this weekend. Hell, nothing worse could have happened! Who else knows about this kidnapping?"

"You, me, the two guards, and, of course, the kidnappers."

"Jake, I'm responsible for this fiasco. I'm going to get her back, before she gets hurt. And before any of our other partners get nervous. So, don't tell anyone else."

"Let's not jump into a hasty rescue," Jake warned. "CMR could be trying to draw you into the open."

"Well, it's working. I couldn't live with myself if I sat back, ignoring Maida's danger."

"Are you crazy? We have only forty-eight hours till you file. Don't jeopardize everything we've worked for. Once you file, Maida is no longer useful to them."

"Jake, don't you understand. I'll never file the claim. Someone from CMR will confront me when I go to the records office. They will threaten to kill Maida, unless I forfeit the claim. There's no way I'll leave her to the whim of those bastards."

"You could lose the gold—and your life."

"She means far more to me than the gold. And if I don't make every attempt to free Maida, we lose anyway. Trust is the adhesive holding this Association together. Sooner or later the other partners would learn we took no action. That would startle them into wondering if any partner would come to their aid, if they too were in trouble. I would be planting a malignancy within the organization I created. Just like a virulent cancer overtakes a body, it would kill the Association."

"Get a grip on yourself, Nolen. We don't even know where she is."

"There's one possible way to find out. In San Francisco, Cholo Cantera mentioned that the CMR men hunting me, were working out of a place in Reno. Maybe, they took her there? If not, someone there might know where she is being held. If that doesn't work, we'll abduct Tower. Then, we can force a trade." He snatched the phone from his desk. "I'll call the FBI agent, to see if he'll divulge that Reno address."

A soft-spoken secretary took Nolen's urgent request and promised to deliver his message. The two men waited by the phone. Nolen paced the floor, incessantly, while Jake cleaned his shotgun and Nolen's pistol. The endless minutes ticked away. Suddenly, the telephone rang, startling both men.

Nolen grabbed the receiver. "Martin here."

"This is Cantera. What's up?" His question was nonchalant, unaware of the urgent nature of their discussion.

"Cholo, I need the address of the place in Reno, the one where you have the wiretap. Apparently, that is the control center for CMR operations. I want my people watching them."

The phone was silent for a moment, then Cantera replied, "That seems reasonable, since you're almost ready to file."

"You're very good at your job. How did you smell that out?"

"Yesterday, our wiretap team monitored a significant increase in telephone traffic at that house in Reno. We pieced together several incoming calls and concluded you are filing soon. Is it true?"

A shudder shook Nolen. Maida must have been forced to tell CMR about the filing. "Yes, Monday afternoon at 3:30 p.m."

"Thanks for telling me. The address you need is 591 Hubera Street, in Sparks. There are a few other things you might want to know, as relayed by your local Gestapo agent."

Nolen wanted to slam down the phone, and race to Reno, but he steeled himself to act calm. "Anything you have to offer would be helpful."

"I received some interesting information from NORAD's Missile Warning Center. The authorization letter from my boss convinced a classmate of mine, assigned to the Denver FBI office, to help. My friend acquired the schedule and observation footprint of all Soviet and Muslim spy satellites that scan the Pacific states. At the time you saw that jet, no foreign satellite was in position to monitor any depot activity."

"That fact supports everything else I've told you."

"Unfortunately, it won't help you much, since the SAC ordered my return to Sacramento. He's assuming direct control of all field operations on this case. So, the next time you see an FBI agent, his mission will be to arrest you."

"Thanks Cholo, you're an honest man. I appreciate all you've done. If everything goes as planned, I'll establish an education fund for Mexican-Americans in L.A."

"I never expected any compensation for helping you."

"I know."

"Nolen, I'm powerless to do anything else to balance the scales of justice. You're on your own."

Twilight rays of sunlight filtered over the blue-green hilltops west of Reno, as Nolen and Jake rolled through the outskirts of Sparks, Nevada. They passed sunburned houses, peeling paint and many vacant lots, choked with scrub weeds. Torn paper, slivered boards, beer bottles, and worn tires littered the shoulders of the roadways. The town favored an ugly step-sister compared to Reno, glittering nearby.

"300 block, 400 block." Nolen verified the house numbers, as their car slowly slid along in the increasing darkness. "There's 521...563. Jake, it must be close...585. There it is! See that duplex, four houses up the street. On my side, with the van in the driveway, like the one Maida's guards described!"

"Double check the address as we drive by," Jake ordered.

"It's blue, by God. The van in the driveway is blue. And, there's Tower's Mercedes Benz parked at the curb."

"You sure?" the security chief challenged.

"I'm positive. Our headlights, plus the porch lamp, were bright enough for me to verify the van color. I also read Tower's ego license plate. We need to bust in right away. She could be hurt bad if we don't reach her soon."

Jake initiated a U-turn and parked the car one block from the house. "You mean dead, and us too, if we make a dumb-ass frontal assault! We don't know yet if Maida is even in that house."

"Jesus, Jake. We don't have all night to debate this dilemma."

"Nolen, we don't know how many people are in there either. Or, what weapons they have. A better idea is to surprise them. Some way, get in close to Maida before they realize what's happening."

"Do you suppose it would surprise them if I knocked on the door and announced I'm a Jehovah Witness?"

Jake smiled, realizing Nolen was instinctively joking to relieve some of the pressure brought about by their situation. "Well, you're not wearing a tie or riding a bicycle. Let's invent something less direct, and less suicidal...like lighting the place on fire, to force them outside."

"Still too dangerous, Jake. If they came outside without Maida, we would be in a real crack. We'd have to fight our way in to find her. And then fight our way back out. How about another approach, possibly setting the van on fire to draw them outside?"

In a rear bedroom of the drab duplex, Maida struggled. She was nude, restrained in a slat-backed, wooden chair. Rope bindings locked her wrists to the sweat stained arms of the chair. Dark blue bruises colored her swollen face.

Boodan, Luke, and two other men watched Tower etch a scarlet tattoo of cigarette burns across Maida's stomach. He was punishing her for failing to answer his questions. She jerked, first left, then right, as the hot tip singed her delicate skin. The smiling Cajun stood behind her, muffling her screams with a rolled bath towel wrapped across her mouth. Slowly, he released the gag allowing her to gasp deep, ragged gulps of air.

She glared at Tower, seated in front of her. There was a strange shimmer in his eyes. He tapped a sharpened ice pick against the top of her left hand. His excited breathing revealed the enjoyment he achieved from inflicting torture upon his vulnerable captive. She shuddered at his blatant vileness.

Maida looked about the room, unable to believe the other three men would casually sit, doing nothing to stop her tormentor. Her eyes begged for help.

Tower signaled to Boodan. He grabbed Maida's stringy, tangled hair. The Cajun forced her head around, so she had to face her torturer. "You know where Martin is," Tower softly whispered. "Quit protecting him."

"No, no. I'm telling you the truth. He's always moving around. That was his intention—for his own protection. I swear, I can't tell you what I don't know!"

Boodan jerked the towel around Maida's mouth, cutting off her oxygen. With wide eyes, she watched Tower slowly raise the ice pick. Suddenly, the salty odor of her urine filled the room.

He slammed the pick down, piercing the flesh and cartilage of her unprotected hand. It vibrated, as blood squirted around the rusty steel shaft. She fainted, unable to cope with the brutality.

Impatiently, Tower nodded to Luke. He hefted a bucket, then sloshed ice water onto Maida's chest. Gradually, she regained consciousness.

"Now, do you have something to tell me?" Tower shouted.

"Yes," she whispered.

Tower leaned forward. "What is it?"

Gathering her remaining strength Maida spat in her molester's flushed face. "Scum!"

Tower snatched a pistol from a side table and jumped on Maida. He struck her until Boodan and another man, wrestled the weapon from him, and dragged him off the woman. Tower yelled, "You dirty bitch! Now, you did it. We're going to gang rape you. You'll never be the same. You'll never want a man to touch you, not after we get finished!"

Boodan and his helper shoved the snarling, struggling executive into the hallway. The Cajun slammed the bedroom door, blocking his view of Maida. "Get in control!" Boodan ordered. "We need her frightened, not in shock, or dead. She can't supply any information while unconscious."

"I don't give a damn!" Tower shouted. "She'll pay the price for spitting on me. She needs discipline."

"She'll get it." Boodan continued pushing Tower down the hallway toward the front door. "But you're startin' to go too far. That jeopardizes our chances of gettin' the information we need. So, I'll finish grillin' her." Boodan opened the front door, then forcefully guided his cousin outside.

Tower abruptly recognized he was standing in the driveway. He ran his hands through his hair, straightening it. He shook himself, muting his emotions. "Tell that whore if she doesn't cooperate, I'll return to punish her."

The Cajun opened the door to the Mercedes Benz. "Sure. Now drive back to your nice hotel. Relax. I'll call you later, when I'm through here."

Boodan wiped sweat from his forehead while he watched Tower depart toward Reno. "That's one weird guy," he muttered.

"Sick is a better description," his helper commented.

Boodan pointed to the van. "Go get us some brews. I need a cold beer, before we start using electricity on the bitch."

33

Fight

Nolen and Jake watched the CMR man enter the blue van, as Tower drove toward Reno. "Looks like we just found our way in, if we can seize that van," Nolen decided.

"Right," Jake agreed, switching on the ignition key. They followed the van for several blocks, until the driver stopped at a brightly-lit liquor store.

Jake parked next to the passenger's side of the van. While the CMR henchman browsed the small shop, Nolen went to the rear of their car to raise the trunk lid. Jake strategically positioned himself at the newspaper stand, adjacent to the store entrance.

Moments later, the unsuspecting kidnaper exited the store, carrying two cases of beer. Jake casually followed him, acting as though he were scanning the headlines of the paper. The man set the beer on the pavement to open the sliding door of the van.

Jake lunged forward. He drove his huge fist into the smaller man's kidney, dropping him to his knees. Nolen hurried from the rear of the car to help toss the gasping man into the van. He closed the sliding door behind the three of them. Jake handcuffed the CMR agent's hands around the base of the rear passenger seat. He gagged him while Nolen returned to the car and quickly backed it into the street. Jake trailed behind in the van, until they found an alley several

317

blocks away. There, Nolen stopped and then hastened into the van.

Jake clicked on the faint interior lights. "Better work fast, in case that clerk saw us jump this guy and called the police."

Nolen pulled the gag from their prisoner's mouth. "OK, asshole. Do you know who I am?"

The CMR man's eyes grew even wider. "You're Martin."

"You win the kewpie doll. I want some answers!" Nolen removed his Swiss knife from the carrying case strapped to his belt. To prove his determination, he brandished the stainless steel blade against the man's cheekbone, slightly beneath his right eye. "Is Maida Collins in the duplex? Yes or No."

The man did not reply until Nolen pressed downward with the razor sharp blade. "Yes, yes! She's there!"

"How many people are in the house guarding her?"

"Only two."

"Doing what?" Nolen again applied pressure with the knife.

"Trying to find out where you and the mine are located."

"What part of the house are they in?"

"Right side of the duplex, right rear bedroom."

"Anyone guarding the front door?"

"No."

"Is it locked?"

"Yes."

"You got a key?"

"It's the brass colored one, on my key ring."

"Listen close. We're returning to the house, and I'm going to go to the front door to use this key. If it doesn't work, I'll kill you. If things inside the duplex aren't the way you say, I'll kill you. Now, is there anything you want to tell me?"

"No, it's just like I said. Look guys, I've got no love for CMR. Just let me go!"

Nolen and Jake briefly discussed their plan, then went to work. They uncuffed the man, stripped off his light jacket, and made certain he was securely bound and gagged. Nolen donned a shoulder holster, put on the jacket, and shoved several items into the pockets of the coat. Meanwhile, Jake retrieved his sawed-off shotgun from the car. The big man verified that it was loaded and the safety was off. Then, he returned to the truck, where he laid on the floor. Nolen hopped into the driver's seat. "You set, Jake?"

"Yeah, let's kick ass."

Minutes later, Nolen pulled onto the cracked cement driveway bordering the duplex. He shoved back the sliding door and grabbed a case of beer. Keeping his face lowered, he moved rapidly to the porch. After verifying no one was peering through the drapes, he set the beer down, then reached up to unscrew the porch light.

When the light blinked off, Jake rolled out of the van. He ran to the house, positioning himself beside the door. He grasped the shotgun barrel in his left hand and braced the butt against his right hip, aiming the weapon so he could fire into the entryway.

Nolen removed the pass key, together with a bottle of tear gas, from his jacket pocket. Holding the gas can in his left hand, he tried the key with his right hand. The key did not fit into the lock. He turned it over and tried once again. It slipped in and turned, allowing him to shove the door inward, as he stepped to one side.

In a half crouch, Jake flowed through the doorway into the unlit house. The shotgun and Jake's head moved from left to right, as if locked together. First, he checked the living room, then the kitchen. No one was visible. Nolen entered behind Jake and quietly closed the door. He dropped the tear gas and key into his jacket pocket and slipped his 357 magnum from his shoulder holster.

A muffled scream sounded from the rear of the building. Nolen nodded toward a hallway where they could see a slash of light escaping from beneath a closed door. Jake crept forward. The stubby barrel swiveled in a silent, deadly arc as they passed the opening of the dark dining room.

Nolen touched Jake's shoulder. He made a jab with his hand, indicating they should move more quickly. After passing several unopened, noiseless rooms, they stopped at the end door. Taunting voices echoed from inside the room, followed by laughter, coupled with Maida's sobbing pleas for mercy.

Jake positioned himself, one foot slightly behind the other, the shotgun at his shoulder. Nolen reached around him and turned the door handle. The big man kicked the door open and jumped into the room, surprising Boodan and another CMR guard. The Cajun froze, his arm raised above Maida's punctured hand. The other torturer dropped the gag covering her mouth, as he shoved his arms above his head.

"Move away from her," Nolen ordered. The startled men complied, stepping back against the wall.

Nolen holstered his pistol, rushed to the chair, and began untying the ropes. "Maida, it's me, Nolen. Hold on. We'll have you away from here in seconds. You're safe now."

Though exhausted and brutalized, Maida tried to raise her head. "Three...three..." she mumbled.

"What is it?" Nolen asked.

Unexpectedly, the shotgun exploded as it was knocked upward by Luke. He had returned from the bathroom and jumped Jake from behind. Instantly, Boodan dove onto Nolen, while the third CMR man leaped toward a 45-caliber pistol, lying upon a side table.

Charging backward, Jake slammed Luke into the hallway wall. Simultaneously, he watched the deadly semiautomatic pistol come off the table and swing in his direction.

Jake lunged sideways as the pistol exploded. Plaster and wood splinters erupted near his head, as the bullet blasted through the sheetrock.

Jake jerked the shotgun trigger, blowing the CMR man from his feet. Twisting around, he drove the butt of the weapon into Luke's stomach. Freeing himself from his attacker's grasp, he smashed a boot into Luke's face.

Jake spun around, spotting Nolen and Boodan thrashing about on the floor, slugging one another. Jerking aside, Nolen snatched the tear gas container out of his jacket pocket. He rammed the bottle against Boodan's chin, while pressing the canister discharge button. Immediately, the one-eyed Cajun was blinded, choking on the burning, itching gas. Nolen kicked himself free of Boodan and slid away from the white vapor. Rising to his knees, he released his pent-up anger by smashing his right fist into Boodan's jaw, dropping him unconscious onto the carpet.

With their eyes tearing from the gas, Nolen carried Maida outside. He placed her gently in the van. Quickly, Jake backed the vehicle into the street. Then, Nolen unlocked the CMR man from the seat, and recuffed his hands behind his back. As the truck sped down the avenue, Nolen yanked his prisoner upright near the open sliding door.

He stuck his face next to the other man's ear. "You lied about how many guards were in the house. I almost got captured, asshole! You're lucky I don't fulfill my promise to kill you." Through the door he launched the man and watched him slam into the pavement and bounce into the side of a nearby building.

At the alley, they switched to their car and then sped toward the Reno Airport. Police sirens began whining in the distance. In the back seat, Nolen wrapped Maida in a blanket, hoping to keep her warm, and prevent her from drifting into shock. Using the first aid kit he had brought, he bandaged her tattered hand and attended her other wounds, as best he could.

"Nolen, I prayed you would rescue...me...somehow find me." She coughed, spitting blood across Nolen's shirt. "After they began torturing me...I...I lost hope." Her terror fueled her ramblings.

"I'm so sorry, Maida."

"Never expected to be...brutalized. Those animals enjoyed hurting...bragged...going to rape me! Keep those men away!" she weakly pleaded.

Nolen gently rocked Maida in his arms. "Honey, you're safe now. We're going to fly to the Susanville hospital. The best doctor in town will care for your hand. Maida, I promise I'll never again put you in a dangerous situation. Guards will constantly be next to you, until I file on Monday. Then, you won't have to worry about CMR ever chasing you again. Please, try to stop shaking. Just rest in my arms."

"We killed that man!" she sobbed. "He was mean—vicious! But...but...it sickened me to see him slaughtered." Tears streaked her swollen cheeks.

Jake replied, "No one's been killed yet."

"Saw you shoot that man. His chest...was a mass...blood everywhere."

Jake removed a shotgun round from his coat pocket and laid it on Maida's lap. "I always ensure my first two chambered rounds are not slugs, when I'm doing in-house combat, where civilians might get hurt. I make these loads myself, filling them with rock salt. The salt rounds effectively knock people down, when you're close to them, in a building. Gives you a chance to decide if you really must kill. I learned the technique doing drug busts. That guy looked a mess, and his wounds will hurt like Hell, but he won't die."

"Nolen, I can't do it any more. Give someone else my job."

"Maida, you don't mean that."

"It's over."

"Honey, we've almost succeeded."

"I don't want to be alone or scared. Too much worry...too much. This isn't worth it."

"You'll feel different after the doctor cares for you. In the morning, you'll see everything in a better perspective."

"Listen to me! I can't...can't do the things you ask. I'm leaving."

Nolen could hardly speak, because he knew he was the catalyst of her injuries. She meant what she said and he knew it. Tears filled his eyes. "Don't leave me," Nolen whispered in her ear. "Please keep me, Maida. You're so important to me. How could I go on without you?"

Maida buried her face against Nolen's chest and cried. Deep sobs poured from the pain in her body and the wound in her soul.

Maida jerked awake in the dark hospital room, when she felt the hand touch her arm. She frantically scanned the moon-lit room.

"I'm sorry," Hilda said. "I didn't mean to alarm you. I came to see how you're doing. Jake told me you were here."

"Are we alone?" A shiver shook Maida's body.

"Just you and me, dear. Don't worry. Jake has two huge guards outside your door. Two others are in the bushes by your window."

Maida forced a shadow of a smile onto her lips. "I'm a little woozy from the pain medication. How's Digger?"

"You're so kind to ask. The doctor is trying some different medicine. Digger's a little better. I guess, I've said that every day since the accident. I can't bear to believe otherwise."

"I'll visit him when I start feeling better."

"Jake said you and Nolen are having troubles?"

"I just don't know how we can continue."

"What made you so upset? Is it because he's not here?"

"He's never here. And because he disappears every time after saying he cares so much for me. Touch and go, together with a few phone calls, don't create a relationship."

"That same anger gnawed at me during the early years of my marriage. For a long time, I couldn't understand Digger."

"I understand Nolen. But my life's dream isn't to file on a vein of gold."

"It's not mine either," Hilda agreed.

"But Nolen told me it was Digger's dream and yours."

"No, dear. All I've ever wanted was to see Digger happy. Like most good men, he strives all the time. I accept him the way he is. It's the price I must pay to hold the wonderful love he gives me. He is a rare man."

"It's so unfair."

"Times like this happen in many women's lives. You're at a crossroads. Wondering if Nolen really loves you, or is he only caring for himself? I can't tell you whether leaving or staying will be best for both of you. But, I can tell you, I think Nolen is much like Digger."

"Which part? The one that keeps running away to prospect? Or, the one that is the caring husband?"

"Did Nolen tell you how Digger made him promise to find the vein? It was just before Digger slipped into the coma."

"I didn't know that."

"Nolen cares for you a great deal, Maida. He's struggling to not turn hard, even though he's been hurt many times by women so very important to him."

"I've felt his hardness, a callousness, a cold withdrawal."

"Once he fulfills his promise, he'll return a love that will make all your regrets seem minor. Give him a chance to love you, child."

"I'm tired of waiting, tired of appreciating the warmth and then living with the lonely chill." Maida closed her eyes, anxious for the drugs to carry her into sleep, far away from her grief.

34

Blocked

27 September

At precisely three o'clock, Nolen parked his car a block from the Recorder's office. Momentarily, he studied the small town scene, but he failed to spot the men waiting for his arrival. He slid from his car, and strode toward the stone steps leading to the state building.

A tall, muscular man wearing a down vest sauntered from a store entrance onto the sidewalk, following the prospector. Near the courthouse steps, another man left the bus stop and also walked toward Nolen. Simultaneously, two men in cowboy hats laughed and shoved each other, while jaywalking across the avenue at an angle which would intercept Nolen.

Nolen noticed the man from the bus stop first. Cholo Cantera blocked his path. "FBI, he announced. "Nolen Martin, you're under arrest."

"I didn't think it would be you," Nolen replied.

The cowboys and the man in the down vest grabbed Nolen, yanking his hands behind his back. As they handcuffed him, a blue sedan screeched to a halt beside the scuffling group. Cantera shoved Nolen into the back seat, slamming the door shut behind them as the car sped away.

An older man, seated in the front passenger seat, watched Cantera search Nolen's pockets. The young agent passed a wallet to his superior.

The older man spoke while thumbing through the identification cards in the wallet. "Allow me to introduce myself. I am the Senior-Agent-in-Charge, Sacramento office of the FBI."

"What are you arresting me for?" Nolen challenged.

"Don't act so surprised, Nolen. May I call you by your first name? I feel I've grown to know you quite well, since we started chasing you seven weeks ago. Ever since you murdered the man in the gas station."

"I didn't murder anyone," Nolen responded indignantly.

"Later, you broke SEC laws and committed several other violations. I can hold you for any of those illegal actions."

"That's a pile of crap. You're just trying to prevent me from filing my claim!"

The SAC smiled, ignoring Nolen's anger. "I have no idea what you mean." Sarcasm oozed from the SAC as he turned away.

"Listen, you liar. I know what the government has been doing at the depot. Stop this car, and get rid of your goons. Or, I'll let them know too!"

The SAC spun around, concerned over Nolen's threat. His prisoner glared steadily into the older man's dark-ringed eyes. "Pull over," the FBI agent ordered. Seconds later, the two antagonists sat alone. "What about the depot, Martin?"

"For several years, the Army has had a secret laboratory functioning there. They've been developing new biological warfare agents. Routinely, they remove the material by jet aircraft, when they know foreign spy satellites are not in position to detect flights in this region."

"What a far-fetched story," the SAC countered. "The United States ceased all research on biological agents in 1969, after President Nixon banned the Department of Defense from such activity. Governments around the world know the US supports the subsequent U.N. conventions, which forbid production of biological agents. No one would ever believe such a preposterous story, coming from a blood-thirsty killer—you lack credibility. You have no proof, just accusations."

"Don't bet on that."

"What do you mean?"

"At my insistence, a pathologist again analyzed a blood sample taken from my foster father following his accident. Would you like to know what the doctor found? Some damn unknown bug, still identifiable in the blood. Something put together from one of the world's worst plague germs. A new organism, probably developed through DNA splicing."

"Surely, you realize we can show that the doctor who performed the pathology is not qualified to identify biological agents."

"The specialist who did the blood work said it was easy to find, given the hints I passed to him. The bug enters the body and is absorbed into the blood stream, where it is distributed to all the muscle groups. There, it attaches to nerve endings, causing the muscles to convulse. The doctor estimated that in large enough doses, it could kill a person within seconds. In lesser doses, it causes incapacitating seizures or heart attacks."

"The Justice Department will tell you that's all speculation—insufficient hard evidence."

Nolen yelled at the FBI agent. "That's what put my foster father in a coma! That dammed germ developed at the Army depot!"

"You're bluffing," the FBI agent blurted, as his stomach began to burn.

"You bastard! Drive me back to the courthouse before 3:30. Otherwise, my lawyer will release that pathology report to TV Station 25 in Oakland. They will present the report as evidence for their feature story on the five o'clock news. The American public will learn clandestine members of our military have been illegally developing biological munitions. They'll further substantiate the story with pictures of Digger, lying in a hospital bed, in a coma, with his crying, elderly wife clutching his hand."

"You cannot connect the military with Digger's coma."

"Wrong. The story will detail the US military motivation, to checkmate a twenty-nine year Russian and Muslim build-up of biological weapons. To corroborate the story, we'll refresh public memory about Russian use of nerve agents in Viet Nam, Cambodia, Africa, and Afghanistan. To spice the story, we'll highlight Iraq's use of germ warfare against Iran in the 1980s and against the US in the 1990s. Make the wrong decision, and tonight the nation will learn how our government exposed its own people to a toxic biological agent."

"Trust me," the SAC argued, "our clout can convince the heads of the major media networks that the story is a fabrication. We can stop the news release."

"So, how will you explain the pathology report my lawyer is prepared to send to the United Nations in New York? The Kazakhstan and Arab extremists will find it interesting reading. Especially, since they don't have an antidote for the new germ. Release of the report will cause the US to lose the technological edge it now enjoys!"

The agent's face turned ashen. "You're a miserable traitor!"

"Far from it. I never meant to interfere with what's going on at the military depot. And I'm willing to remain silent about any sensitive information, if, and only if, you notify whoever is running this show, to back off. No trumped up murder charges. No SEC investigation. No IRS audits. No federal or state challenges or delays to my mining claim. No trouble of any type. Or, I'll let the world know about the Army's nasty killer germ."

"You're a blackmailer."

"Once I am satisfied that the government won't interfere, I'll relinquish all my evidence. Now, release me. Otherwise, you alone will be held accountable for national security information reaching our enemies."

Minutes later, Nolen stood in front of the Recorder's office, watching the FBI auto drive away. Roger Benjamin waited for him at the top of the steps leading into the county building. "You're late. Who were those men?"

"No time to explain." Nolen breathed a sigh of relief, hurrying along the walkway. "I'll tell you later."

A half block away, in a large travel trailer, Boodan and several men observed Nolen's return. "This is crazy," Luke grunted. "That bastard gets snatched. Then, a little while later, they put him back on the street."

"Shut up!" Boodan hissed. "We need it quiet to hear what's being said."

Inside the building, the senior clerk paced. She recognized Nolen as the door opened. Unobtrusively, she flipped a hidden switch controlling a microphone taped to her side of the opaque glass counter divider. She verified the active light shone bright, even though she had checked the system four times in the past thirty minutes. "Can I help you?" she asked, anxious to sound calm.

Nolen took a deep breath. "Yes, I'd like to file a claim on a mineral strike."

"What type of deposit is it?"

"Gold."

"Well, you are a lucky man." She reached for a claim form. "Do you have a mineral sample available from the site?"

"Here's a piece of gold taken from the vein, as well as a copy of the assay."

"Have you marked the site boundaries?"

"Yes, I used six stone discovery markers, with a can buried under each one. Inside each can is a paper with my name and all the rest of the necessary site description information."

"What size is this claim?"

"1500 feet long by 300 feet wide."

"Where is the location of the center of your strike?"

"It's at 29253771, as shown on the Milford quadrangle, Department of the Interior map."

"H'm, that's on national resource land, so you will be filing under..."

Nolen finished the sentence. "The US Mining Act of 1872 on unreserved land."

"Seems like you know what you're doing."

"I've followed the procedure many times."

"Well, everything appears in order. Now, I must verify some records before I can accept this as a valid claim. It will just take a few minutes. Why don't you folks take a seat? I'll call you when I'm finished."

In the travel trailer, Boodan radioed a message. "Helicopter teams, site coordinates are 29253771."

"Team One here. I've got a good copy."

"Team Two, roger."

"Team One," Boodan ordered, "do an aerial search to locate six boundary markers. Land and build a seventh one."

"We're moving now," came the reply from the pilot.

"Team Two, after Team One notifies you that they have completed their task, fly the Susanville mayor to the site. He must witness you openin' the can under the seventh marker, with our claim data in it. Contact me when you have finished!"

Seated next to Boodan, a bespectacled man scribbled several entries into Digger Jorgenson's prospecting notebook. Carefully, he focused a heat lamp onto the page. "Hurry up, damn it," the one-eyed Cajun growled at the man.

The wrinkled writing specialist inspected his recent work. Snapping off the lamp, he handed over the notebook. "Here you are, Boss. Pretty as you please. No one will be able to detect that the last page is a forgery. I practiced Jorgenson's scrawl for hours, until I got it exact. Besides, I even used the same type of ink that Jorgenson used for those original entries."

With his good eye, Boodan verified that the new entries appeared to match Digger Jorgenson's handwriting. Then, he tossed the book into a leather attaché case and stepped outside. At the rear of the trailer, he waited for Tower to help Hilda Jorgenson from his Mercedes. Boodan handed his cousin the attaché case and said, "It's all set."

"Good. Remain in the trailer, and tape record everything that is said."

"You sure you won't need me?"

"I'll be fine. I've been waiting for this opportunity for a long, long time. Come along, Hilda. We must hurry."

A Sheriff's patrol car rolled to a stop next to the state building, as the mining executive and the old woman crossed the street. A brown-haired deputy exited the vehicle to begin inspecting an illegally parked car. He looked up as Hilda, assisted by Tower, hobbled past.

Minutes later, confrontation noises echoed from inside the courthouse. The deputies rushed into the Recorder's office where they found Nolen and Tower, arguing. Roger Benjamin stood as a barrier between the men, restraining them from reaching each other. Hilda Jorgenson sat on a nearby bench, weeping.

The brown-haired policeman and his partner separated the two men. "All right!" the cop bellowed. He nodded toward the female clerk. "You, tell me what happened. Everyone else, keep quiet."

"Well, this young fellow came in about fifteen minutes ago. He wanted to file a mining claim. Later, that man, in the three-piece suit, and the older woman, came in. The well-dressed man began calling the younger one a thief. That caused the yelling. The young one hit the other guy, and everyone tried to stop them. Then you came through the door."

The officer looked at Tower. "What's your story?"

"I'm here to prevent this man from stealing an old woman's mining claim!"

"That's a damn lie!" Nolen shouted.

"Shut up," the deputy snapped. "You'll get your turn."

The executive continued. "Sir, my name is T.J. Tower. I'm Vice-president of Continental Mining and Refining, in charge of the Exploration Division. Earlier this year, a friend of mine, Mr. Digger Jorgenson, signed a contract, agreeing to work for my company. He was prospecting for us, when he was injured in an auto accident."

"Get to the point," the officer ordered.

"Unfortunately, Digger is this thief's foster father. Martin knew Digger had a good lead on a mineral discovery. So, he started hunting for it, after Digger was hospitalized. Eventually, he located the mine. But I have proof Digger Jorgenson found the strike before he became debilitated. Legally, that find belongs to his sick widow. She's seated over there, officer." He pointed to Hilda who was still weeping.

"This is a huge lie," Nolen objected.

"A week ago," Tower continued, "Mrs. Jorgenson spoke with me. She advised me that her husband often hid his prospecting notebook in his truck. Fortunately, this morning, I located his wreck in a junkyard near Susanville. There I found Digger's notebook with a dated assay report."

Tower opened his attaché case and extracted several documents. "Officer, here is proof—the notebook, the assay report, the signed contract. The final written pages in the book list the coordinates of the strike, as well as notes detailing the discovery. Notice the date on the assay report is eight weeks old."

"This story is complete fantasy!" Nolen blurted.

"Shut up, pal." The policeman paused several moments to review the items, then nodded. "OK, continue."

"We went to the site earlier today. We found Nolen had been there, and had replaced claim data under six of Digger's discovery markers. But he is not as clever as he thinks. He missed the seventh marker. I rushed back here just in time and caught him!"

Glaring at Nolen, Tower asserted, "This man's attempting to steal the claim which rightfully belongs to his foster mother. Perhaps I was wrong, officer. When I saw him standing there, I just couldn't control myself. I started yelling at him, and then he hit me." Tower began rubbing his cheek which was beginning to swell.

The officer turned to Nolen. "Go ahead, let's see if your story can top this one."

"The Recorder told you the truth. Tower didn't. Yes, I found the strike a month and a half ago. But it took me seven weeks to get everything together. That's why I'm filing now. That contract is a forgery. The assay report is too. And so are those notebook entries. CMR is the culprit attempting to steal the claim from Hilda, not me."

"That's not totally correct," the Recorder interjected.

Nolen stared at the woman. A stunned expression flashed across his face. "What do you mean?"

"CMR filed a notarized grubstake contract with us in mid-July. Several weeks later, Mr. Jorgenson was in here searching through our records. He said he was exploring for CMR. I can verify the contract is valid."

Nolen yelled at the Recorder, "No matter what anyone says. it's all lies. Tower, you can't take my claim away. It's filed. You can't stop me!"

"It's not filed yet," the Recorder countered. "I haven't notarized the claim affidavit. By state and federal law, I possess the authority to decide who to accept on disputed discoveries. In this case, I'm choosing CMR, based upon the substantial proof showing Mr. Tower's representative located the site prior to Mr. Martin."

Nolen leaped at the clerk. The two officers restrained him, before he could reach the startled woman. One policeman tripped Nolen, shoving him to the floor. The other yanked his arms behind his back to handcuff him. Their manner was rough and decisive.

After the officers dragged Nolen from the office, Tower verified that the clerk notarized the claim for CMR. Gloating over his accomplishment, he used his cellular telephone to call his company Chief Executive Officer. Tower informed his Japanese CEO that the Exploration Division had just secured the gold strike. It was a victory, for which he proudly accepted all credit.

I'm invincible, Tower thought, as he tossed his phone into his attaché case. Totally brilliant. Finally, the Presidency of the western division is mine—at double my salary. Never again will they dare criticize me! From now on, everyone will have to kiss my ass.

35

Filing

The police cruiser scooted away from the courthouse, as Tower emerged from the building. The executive displayed an obscene gesture toward Nolen. "I hope you rot in jail!"

The brown-haired officer turned in the front seat to speak through the protective grill. "Either that guy doesn't like you, or he's putting on a good act."

"He doesn't like anyone," Nolen answered.

"Are you OK? You put up a tougher struggle than we expected."

"Yeah, I'm fine. You did great. But, these handcuffs are killing me."

The patrol car traveled several more blocks before parking. Jake Samson, dressed as a local sheriff, stepped from the sidewalk to join Nolen in the back seat. He removed Nolen's metal shackles while saying, "CMR will be gone in just a few moments."

Nolen massaged his wrists, rearranged his clothing, and combed his hair. "Good. Things seem to be progressing, as we planned."

The men waited almost ten minutes, then returned to the county building. As the patrol car again parked outside the Recorder's office, a TV van arrived. Cameramen began unloading equipment. Pam Pendleton met Nolen, when he stepped from the police cruiser.

"You ready to become a celebrity?" he asked her.

"Yes, half an hour ago we verified the satellite link, as well as our camera functions. The lead-in is already taped. Both my primary and backup men are prepared to film."

"OK. Jake goes in first, I'll follow, then your crew. Start filming just before I pass through the door."

When Jake entered the Recorder's office, Benjamin was talking on a pay phone. Jake approached the senior clerk's desk. He started discussing the recent incident. Suddenly, Nolen burst into the room.

"What do you want?" the startled Recorder inquired.

"I want to file a mining claim," Nolen replied calmly.

Puzzled, the clerk turned to Jake. "Officer, we went through this matter, a few minutes ago. Two other deputies arrested this man shortly before you arrived."

"No, they merely removed him from the scene, so there would be no further trouble. Once the others left, my men released Martin. The ruckus is over now, and the claim they are disputing is a civil matter which can be resolved in court. No need for anyone to go to jail."

The clerk strutted to the counter. "Well, you can't file for the CMR site. I've already recorded their claim."

"That's fine. I want to file on a different location." Nolen handed her a filing form he had previously prepared. "Everything is correct. There are two portions of land which follow the vein. Both are 1,500 feet long by 300 feet wide. I have listed the claim names, locations, number of acreage feet, section, township, range, and base meridians. The claim lies on Bureau of Land Management, unappropriated, public land. Try county book 271, page 82. One of my attorneys inspected your records two weeks ago. Therefore, I'm certain about the status of the site."

The Recorder's pallor became chalky. She focused on the news team standing in the doorway, then yelled, "Why are you filming? This matter is of no consequence to the press.

Get out of here!" She reached for the hidden microphone switch.

Jake's hand moved far faster than hers. "This man has a legal right to petition for a new claim. If you do not accept his filing, it would appear that as a state employee, you're illegally collaborating with CMR. In fact, this device seems to be evidence substantiating that alliance." He ripped the microphone from the glass partition. "If you don't process this man's application, I'll confiscate the news crew's film. A grand jury would be most interested in viewing how you handle this filing."

The perplexed Recorder dropped the filing form, as if it were a deadly snake. She contemplated for several seconds what action to follow, then shook her head while turning toward the document shelves. She located several county record books, then notarized Nolen's claim without uttering another objection.

Nolen verified that she had not deliberately entered, or left blank, any entry which would invalidate his claim. Smiling at her, he paid the required administration fee, being certain to collect his receipt.

Outside, a squat, red and silver armored truck rumbled to a halt. The TV crew flashed white lights onto the rear doors of the truck. Nearby townspeople became attracted by the unusual activity in front of the public building. Rapidly, an inquisitive crowd formed in the street surrounding the TV cameramen.

Hamilton Roberts assisted Hilda from the cab. Adjusting his toupee, he hastened to meet Nolen and Pam behind the vehicle. Two uniformed security guards, carrying shotguns, opened the rear doors of the truck. A black velvet cloth covered a large mound stacked on the truck floor. The scene heightened the interest of those who gathered around the truck.

Pam urged the milling crowd to be silent, then proceeded to interview Nolen. "Mr. Martin, I understand that you wish to make a sensational announcement."

Before answering, Nolen smiled at Hilda, then began his practiced statement. "Yes, I have discovered the richest gold mine in California history!" Nolen whipped the velvet cover out of the truck compartment. "I located a gold vein three-quarters of a mile long, which assayed eighty-five percent pure. There is a fortune buried in the mountains east of here!"

A gasp echoed from the men and women congregated in the street.

He lifted two large gold rocks above his head, so the crowd could appreciate the richness of his find. The chunks glittered, reflecting the bright illumination of the TV spotlights. A steady murmur rippled through the awed crowd. The chatter became increasingly louder. Bits and pieces of conversations jumped out of the mass of people, which seemed to expand in size by the minute.

"Gold! A gold strike near our town!"

"That's right," Nolen replied. "My discovery is huge. Possibly, there are many more gold veins to be found near my site. This strike is larger than any discovery during the gold rush of 1849!"

"Come off it," one man jeered. "This is just some kind of publicity stunt. I don't believe you."

"It's no stunt," Hamilton responded. "Friends, most of you know me. I manage First Mountain bank in Susanville. Over the years, I've dealt honestly with many of you. Let me assure you, several hundred pounds of ore taken from the discovery site are stored in the vault at my bank. The ore has been assayed and I can confirm that it's nearly pure gold."

Another man within the jostling crowd exclaimed, "Look at the size of those rocks! That's a treasure he's holding in his hands!"

"Where did you find it?" a woman inquired. The film crew swung their cameras toward the throng. People pushed and shoved, as they attempted to gather nearer to the gold, desiring further details from Nolen and Hamilton.

Nolen answered, "I uncovered the vein near Hot Springs Peak, in the Amedee mountains."

Another voice pierced the clamor. "That's within twenty miles of here. My God, imagine if we found part of that vein."

Suddenly, the crowd erupted. People scattered in all directions, scurrying to reach their vehicles. Several women were knocked to the ground and trampled. Jake snatched Hilda from the path of the mob. One of the guards fired his shotgun into the air, warning people away from the armored truck.

Abruptly, from across the street, a station wagon backed from its parking space. Slamming into the armored truck, it showered plastic and glass over onlookers in the street. The excited driver then gunned the station wagon down the road, forcing men and women to leap aside. Nolen tossed his gold into the van. One of the security guards slammed the rear doors shut.

Pam stepped in front of the camera and commented into her microphone, as the sea of pandemonium surged around her. "Ladies and gentlemen, a modern day gold rush erupted today, in Susanville, California, at 4:20 p.m. Men and women are racing to claim their fortune, hidden in the hills northeast of this once sleepy, all-American city. Please stand by. Each hour, I will provide further exclusive Channel 25 news coverage concerning the winners and losers, the lucky and the unlucky. This is Pam Pendleton reporting. Good bye for now, and good hunting."

When the video lights dimmed, Pam yelled to her crew chief, "Transmit my story to the main office while I gather more interviews. Tell the home office that we'll have another story, camera-ready, in fifteen minutes." Pam hurried down the street, forcing her cameraman to awkwardly jog behind her at the end of the thirty foot microphone cable. The reporter knelt and shoved the mike in the face of a bruised and bleeding teenage girl seated on the curb. The girl clutched her torn dress so as not to expose her virgin breast. She cried and babbled, as she tried to answer Pam's questions.

Benjamin jerked Nolen by the arm. "You son of a bitch, tell me what's happening. I thought we lost our strike to CMR?"

"Take it easy, Benjamin. Lucky for us, an honest FBI agent gave me a tip. His wire tap team learned the County Recorder was on the CMR payroll. Hearing that, I knew they were scheming to steal our mine when I filed. So, I had one of Jake's men scout the office. He spotted the hidden transmitter taped to the counter. At that point, it was an easy guess how CMR would try to block my claim during the filing. Consequently, the first set of coordinates I gave to the clerk was false. Obviously, CMR forged those coordinates on the documents they presented. Both Tower and I should be given Oscars for our outstanding performances."

"What about your foster mother, telling CMR where to find her husband's notebook?"

"Half of the majority ownership of the mine belongs to my foster parents. Initially, I felt it wise not to tell Hilda what I had discovered. I feared CMR would contact her. She would only be safe then, if they believed she didn't have any knowledge of the strike. Hilda, why don't you tell him the rest?"

"CMR did contact me," she continued. "They claimed Nolen was cheating me and also cheating Digger. I was so confused, until twelve days ago when Nolen phoned. That's when he explained how he was protecting me. He revealed that he had Jake plant the notebook in Digger's wrecked truck, so we could dupe Tower into filing on the wrong claim. I then informed Tower that the notebook might be found in the pickup. They could not resist the bait. I sure enjoyed my part in misleading CMR for the Association."

Nolen hugged her. "Your crying was perfect."

"I tried my best to play the role of the betrayed mother."

Benjamin growled. "Who the hell are these policemen?"

"The big one, that's Jake Sampson, my Security Chief. Before he joined the Association, he was on the Susanville police force. That's where the uniforms, the cruiser, and two additional men came from. It was imperative to make the CMR crew think the police had me in custody. Jake's friends helped provide the necessary realism."

"I'll be damned," Benjamin exclaimed. "You made CMR play all their trump cards, just to steal a false strike."

"No doubt, Tower thought he had won when I was dragged, struggling from the Recorder's office. Now, there's no way CMR can legally challenge us."

"Nolen, are you always such a slippery con-man?" Benjamin wondered, slapping his partner on the back.

"Save your questions. We still have urgent business to complete. Let's get to our helicopters, and fly to the vein. Kemper already has our convoy of men and equipment moving toward the site. We must arrive before the entire population of Susanville reaches those hills. Jake will drive us to the helicopters stationed at the local high school."

Tower's Mercedes skidded to a stop next to the armored truck, as Nolen and the other Association members were departing. The furious executive jumped from his car. Waving his fist, he stormed toward Nolen. "The County Recorder called me on my car phone. You think you have outmaneuvered me, don't you?"

Jake moved in the direction of the angry executive. Nolen placed his hand on the big man's arm, preventing his protective advance. Facing Tower, Nolen replied, "Yes, and I enjoyed blocking your attempts to swindle my family and my partners. You've been a pirate for years. Far too many people have been steamrolled by you and your company. For once, the little guys retaliated."

Hilda stepped from behind Nolen. Without warning, she kicked Tower between the legs. The surprised group watched Tower's eyes bulge, as he grabbed his groin, falling to his knees in pain. "That's for stealing our first mine." Hilda began clubbing Tower's head with her purse. "This is for trying to cheat us again!"

After Tower's head began oozing blood, Nolen signaled to Jake to stop Hilda's assault.

Tower painfully crawled back into his car. As he drove away, he shouted, "Don't think you're safe yet! I'll get you! No false claim is going to provide you, your bitch girlfriend, or your family with any protection."

Jake turned to Nolen. "It might be a good idea for me to take that idiot out of action for awhile. Sounds like he meant what he said."

"Don't worry," Nolen answered. "Being outsmarted and losing millions will make him very unpopular at CMR. Let his bosses handle him. He's simply an angry, insignificant thief, who has suffered a severe defeat. We've won. Forget him. Let's get to our vein."

Tower was obsessed with his loss. As he drove the few miles to the airport, he feverishly tried to devise his revenge. *That dirty bastard. I must do something quickly to salvage my position within CMR.*

Tower was concentrating so intently that he hardly noticed the many pedestrians running about, cars careening by, and the few police attempting to control the mayhem. An elderly couple, desperately trying to flee the street, almost became his victims.

First, my prospecting teams will occupy the land surrounding Martin's strike. We'll stake claims to those areas for

CMR. I'll advise the company CEO that we control part of the big vein. He will be preoccupied for several days reacting to the reverberations in the gold industry, as everyone learns about the massive discovery. He won't realize the truth until after I recover from Martin's trick. I have time! And I intend to use it wisely.

Screeching his car to a halt at the airport, the executive found Boodan waiting for him. Tower jumped from his car. "How would you like to earn a $50,000 bonus?"

"I don't follow you," the Cajun replied.

"We still have one remaining opportunity to gain control of Martin's mine. I want you to kill him—tonight—tomorrow—or the next day, at the latest."

A wicked smile cracked Boodan's face. This was the type of bonus he lived for.

36

Excavation

27 September

The huge, three-prong ripping plow, at the rear of the tractor, tore into the ridge, seeking the valuable vein below. Periodically, the dozer crashed into an underground rock slab that it could not shatter, jerking the powerful vehicle to a sudden stop. Then, the metal beast would raise its plow and lumber past the hard obstacle. Behind the bulldozer, a mobile crane and two tractors with front-end scoops gobbled bucket loads of loosened soil, which they loaded into a line of waiting dump trucks.

Meanwhile, a cable truck traversed through the tent area. It reeled out a mile-long electrical power cable, bouncing parallel to the upper edge of the excavation line. Following along the cable, a light team lifted the line onto wood tripods. Then, they connected spotlights into the power cable.

When the reeling concluded, electricians attached the main cable into the electrical buss at the rear of a large generator. Shortly after sunset, the generator rumbled to life. Beams of white light illuminated the ridge, supplementing the headlights of the excavation vehicles.

At the base of the ridge, a second massive bulldozer completed leveling a large parking area. An assortment of unloaded busses, trucks, and flatbed trailers were already waiting for relief from the crowded ridgeline. In one corner of the parking area, a Quonset shaped tent was being erected. Machinery or vehicles requiring maintenance during the coming weeks would be serviced inside the tent.

Several hundred yards down the road from the parking area, a third bulldozer finished creating a U-shaped dirt berm, where two trailers containing explosives could soon be parked. The yellow bulldozer spun around, then rumbled uphill to join the excavation effort.

The remaining activity along the ridge was near the mouth of the defile. There, a well-drilling rig bore into the earth in pursuit of water.

By 9:00 p.m., a solid strip of granite had been exposed across the width of the ridge, blocking the steel plows and dozer blades from penetrating nearer to the lode. Only drilling and blasting would allow Nolen's team to search more deeply.

Since most of the crew had completed their initial tasks, they began their primary job as rock drillers. First, they unloaded six motorized compressor drills and hauled fifty jack hammers from flatbed trailers. They uniformly spread the equipment along the exposed rock. Then, fifty-six men donned face masks, goggles, and ear guards to protect themselves from floating, glass-sharp granite dust, as well as from the inevitable jarring noise.

The racket from the generators, drills, and jackhammers grew from a coughing clamor into a deafening thunder, forcing the bystanders to also don ear protectors. Within minutes, a gray fog of rock dust enveloped the tired, sweating men. One by one, they relinquished their jack hammers to relief replacements, who continued drilling the long blast pattern.

The pattern consisted of four parallel rows of 100 holes. Each hole was three feet apart and twenty-four inches deep. Every hole was sunk, pointing down slope, at a thirty-three degree angle, penetrating the resisting rock. The holes were designed to channel an explosive blast, so it would kick most of the stubborn stone downhill from the vein. Kemper had devised this procedure to speed the removal of guarding rock, and allow the waves of workers to quickly return to drilling nearer to the valuable gold vein.

Gradually, the jack hammers fell silent as they reached the proper drilling depth. When the last drill withdrew uphill from the tattoo streak of 400 holes, the dynamite crew cautiously drove their truck forward. They played out a line of white detonation cord. Every ten yards, they carefully placed boxes of explosives on the ground. For an hour, the master blaster supervised the 40 men, gingerly packing the high-powered explosives into the rock. Finally, after walking the entire length of the blast line for one last time, he signaled that he was ready. Immediately, the light crew ran along the ridge, laying down the floodlamp tripods, causing a shallow blanket of white fog to cover the blast area.

Uphill, along the edge of the tent area, Nolen, Kemper, Benjamin, and the balance of the excavation crew lay in the darkness, waiting. They listened while the master blaster verified all men were safely away from the detonation area. Their pulses quickened when the master blaster chanted: "Fire in the hole—Fire in the hole—Fire in the hole!"

A heartbeat later, the earth slammed the men in their stomachs. A thunderous roar stunned them. Tons of rock shot skyward. The watchers, shielding their heads, cringed. Thousands of rock chips returned to earth through the dust-choked air, peppering the ground around the men, like a spontaneous rain shower. Curses pierced the darkness in reflex response to the biting sting of rock pellets.

Numbed by the tremendous explosion, the men waited for their bodies to recover from the vibrating effects of the shock wave. Few were in any hurry to move. Eventually, the light crew repositioned the battered metal floodlights onto their wood tripods. They began replacing the many broken light bulbs. Still, the granite strip seemed barely illuminated, due to the throat-choking dust cloud hovering over the ridge.

Nolen, Kemper, Benjamin and the master blaster stumbled along the length of the blast line, surveying how well the explosives and gravity had succeeded in removing the heavy rock. Three feet of solid stone, some 115 tons of material, had been gouged from the earth and tossed downhill. No obvious ore appeared. The blasting procedure would have to be repeated, until a V-shaped trench exposed the vein. Then, a less brutal method could be used to extract the valuable mineral. Kemper directed bulldozers, cranes, and front loaders to move forward, to begin transferring the rock debris into the dump trucks.

As if prompted by the explosion, Benjamin's impatience erupted. He began to pressure Nolen. "Look, we've been standing around here long enough. Apparently, it will take a day, possibly two, before this excavation crew reaches the most valuable ore. Let's return to the boomtown. My geologists have verified the purity of the gold taken from the face of the vein. I'm tired of waiting for you to sign over my share of the mine!"

Nolen wearily shook his head. "Sure. It's time to take care of you." The two men trudged uphill to their helicopter, waiting above the tent city. They flew to the headquarters building. Together with their lawyers, they formally finalized the agreed-to exchange of mine shares for mining company stock.

As Benjamin slipped the newly acquired mining stock into his briefcase, he commented, "Partner, we should celebrate with a drink, then turn in for a good night's sleep!"

"There is one more item to settle prior to any celebrating."

"What's that?" Benjamin seemed perplexed.

In answer, Nolen opened a side door leading to the conference room. The other board members of Benjamin

Mining entered the room followed by Hamilton, Karl Kemper, and Jake, together with several security personnel.

Surprised by the presence of the other two major stockholders of the mining company, Benjamin growled, "What's going on?"

"I'm calling an emergency meeting of the mining company. The purpose is to elect Karl president," Nolen stated.

"What? You can't do that!" Benjamin objected.

"Yes, I can. Incorporation by-laws allow any stockholder controlling fifteen percent of the voting stock to call an emergency meeting. You just transferred fifteen percent of the stock to me. So I'm calling this meeting. It is in session now."

"This is a waste of time," Benjamin snapped. "I run this company. Furthermore, throughout my lifetime, I intend to remain president. The by-laws also state an additional rule. During emergency meetings, the President can vote proxy stock, as he sees fit, for any stockholders who haven't declared their voting intention and are absent. Since I don't see any of those people, who account for forty-four percent of the voting stock, I'm going to cast their votes to retain the present Board of Directors."

"True, you currently have that right," Nolen confidently answered.

Benjamin warily looked at Nolen. "Now, add in the thirty-one percent I and the other two board members own, then clearly, I control a majority of voting stock. That allows me to dictate the outcome of any issue brought before this meeting. So let's just stop this shit, right now!"

"Unfortunately for you, I won't because of one small detail. You don't control a majority of the stock."

Benjamin slammed his fist onto the table. Glaring at Nolen, he snarled, "OK, wise guy. Prove it to me—and to my company attorney."

Nolen looked toward Kemper. "How do you vote?"

Kemper smiled at Benjamin. "I vote to replace the President, of course."

"That creep's measly five percent, plus the fifteen I just signed over to you, falls far, far short of a fifty-one percent majority."

"Yes," Nolen answered, as he accepted a brief case from Hamilton. "However, over the last month I've acquired the rights to vote twenty additional percent, from others around the state who have recently purchased company stock. Also, during the last twenty-four hours, while you were in the field with me, my attorneys visited ten current stockholders. We informed them that I wanted a new president. Several of those stockholders agreed, after viewing the recent TV reports, just how beneficial it would be if the discoverer of the gold vein were confident with the mining company president. That maneuver allowed the Gold Association to gain control of an additional seven percent of the voting stock."

Nolen removed a stack of documents from the briefcase and spread them in front of Benjamin's lawyer. "Each notarized proxy contains a guaranteed signature which authenticates its legality. An additional statement is attached, granting me specific authority to vote the stock as I so elect."

"You still fall short, asshole," Benjamin snarled. "Your total only adds to forty-seven percent!"

"Which brings us to this." Nolen took a cassette tape from his briefcase and tossed it onto the table. "During our first meeting, I recorded our conversation. I'm sure you recall, I originally offered your company six percent ownership of the gold mine. Six percent, which should have been equitably split among all three board members of your company. However, you were too greedy, unwilling to share any of the wealth from the gold vein with your partners."

"You piece of shit!" Benjamin's face reddened with anger.

"While I filed the claim this afternoon, my lawyer played the tape for your two partners. He then offered the other two board members two percent ownership in the vein. Learning how you screwed them, convinced your partners to exchange half of their mining company stock for partial ownership in the gold vein. Furthermore, they agreed to vote in support of Karl's election. That brings my total to sixty-seven percent of the mining company voting stock. The decision is final—you're out. Karl is the new president."

Benjamin snatched a handful of proxies off the table, scrutinizing them in disbelief. "Are these legal?" he gasped, turning to his attorney for support.

Benjamin's anxiety became vivid when his advisor nodded his head in the affirmative. "You dirty bastard!" Benjamin screamed. He leaped across the table, grabbing Nolen by the throat.

Jake and the security guards instinctively reacted to restrain the kicking and spitting ex-president. Hamilton helped Nolen to his feet, as he coughed and rubbed his bruised throat. Nolen jerked his thumb toward the conference room door and croaked, "Jake, return Benjamin and his attorney to Reno."

"I'll get a restraining order, you son-of-a-bitch!" Benjamin bellowed. "I know judges who can stop you. And I'll make sure you never deliver the first load with 1,000 ounces of gold."

37

Gold Rush

30 September

The film chief positioned his TV camera crew so they could focus on Pam, standing in front of the excavation site. The chief silhouetted her with the nighttime activity bustling across the hillside. He shouted at the onlookers clustered around the newswoman. "Everybody stand back! We go national in ten seconds...five, four, three, two." He stabbed his finger at the reporter.

"Good evening, this is Pam Pendleton featuring another of my exclusive Channel 25 live action reports. It is 9:05 p.m., three days and five hours since the incredible announcement in Susanville. Gold! A wondrous amount of gold, discovered in this part of the state! That revelation has enticed thousands to scramble into the hills straddling the California-Nevada border.

"Seventy-two hours ago, I showed you the ledge where Nolen Martin first viewed his fabulous gold vein. Tonight, I've returned to that discovery site. As you can see, around-the-clock excavation has dramatically altered this hillside, behind me."

The television camera panned along a wide trench cut into the hillside, illuminated by clusters of floodlights and surrounded by constantly active equipment. Workers, controlling jackhammers, chiseled away at the material in the bottom of a large trench. White clouds of dust puffed into the air about them.

"Standing beside me is Mr. Karl Kemper, President of Benjamin Mining Company. Mr. Kemper, would you describe the operation taking place tonight?"

"Nolen Martin hired my firm to mine, transport, and refine the metal. On the day the gold discovery was announced, we hauled in heavy equipment and began clearing away topsoil and bedrock to gain access to the deposit. Ten hours ago, we reached the main seam and started removing high-grade ore."

"Could you describe the size of the vein for our viewers?"

"Sure." He pointed into the 200-foot-long, V-shaped trench. "Here, it's about twenty feet below the surface and surrounded by white quartz. Core samples indicate this section averages eight-feet wide and varies from six-inches to twenty-inches thick."

"How far do you believe the vein extends?"

"That's difficult to say. A deposit can disappear within a few yards. Or, like many South African mines, it may run for miles into the earth. We'll be in a better position to determine that answer after further digging. However, I have inspected the first ten feet of the seam that we've cut into. I'd venture to say, thousands of years ago, a thick, molten stream of gold pushed upward from the earth's core, into this mountain range. So, I predict the vein is big and runs deep into the earth."

"Mr. Kemper, I notice one of your men, uphill from the excavation, is welding a lid onto a large metal box, the size of a truck-portable garbage bin."

"Yes, it contains our first load of high grade ore. It will depart momentarily for our nearest smelter located in Riverside, California."

"What is the estimated value of this first shipment?"

"Well, I'm not sure how low the world spot price has dropped. Conservatively, I estimate around ten million dollars."

"It's a long drive to Riverside. Are you afraid someone will steal your treasure?"

"Yes, Pam, I am. So, we're not transporting by truck. Do you hear that?"

The reporter became aware of a distinct and rhythmic whop, whop, whop sound. Seconds later, a forty foot long, dual-engine helicopter descended out of the blackness. It hovered above the large rust-brown cube.

Beneath the helicopter, the gale force downdraft pounded two men, standing atop the container. They held a metal doughnut ring, which was connected to the four corners of the big box, by thick cables. A third man used red batons to direct the pilot in lowering the helicopter. When it was close enough, the two men raised the ring and slapped it into a carrying hook, dangling from the belly of the aircraft. They then jumped to the ground and sprinted from under the roaring helicopter. Gradually, the cables stiffened and the heavy load swung off the ground, disappearing into the night sky.

"Isn't it a bit unusual to transfer ore by air, Mr. Kemper?"

"This is an unusually rich strike, demanding unusually secure means of transport. Besides, I want a sample of the ore and quartz analyzed immediately. We know the main vein is extremely pure. However we don't know how much gold is contained within the material surrounding it. We'll learn that fact in a few hours, when the three tons of extracted material is processed."

"Thank you, Mr. Kemper." Pam faced the camera again. "This is Pam Pendleton reporting live from Lassen county, following the California gold rush. Remember, you viewed today's most exciting news breaking first, and in depth, on Channel 25. Until tomorrow morning when I again update you concerning this fabulously rich gold discovery—good luck and good hunting."

The bright television lights blinked off. Pam's crew chief handed her a head set, and said, "The station manager wants to talk with you."

Pam conversed, while watching Nolen trudge uphill from the work site to a small helicopter. He waited there for her. When she completed the call, Pam ran to the aircraft, and climbed in beside Nolen, as the rotors began to twirl.

"Nolen, I've got great news!"

"Yeah," he acknowledged in a weary voice. "What is it?"

"Our Nielson ratings are incredible. Each one of my newscasts has had fifty-five percent of the country watching. And, the percentage continues to grow daily. Our competitor stations have been limited to showing crowds of people leaving the major cities. The quality of their coverage can't compare with my continual updates and personal interviews, at the excavation site. Competitor crews are investigating why I can develop such exclusive reports. My God, this thrill is greater than I expected!"

The helicopter shuddered, then leaped into the air, swinging toward the southeast.

"Nolen, my manager insists that we do a film-essay tomorrow morning at the smelter site, to detail the value of the vein. I have already told Kemper, to ensure that no one at the smelter provides interviews to anyone except me. Then, my boss wants another story about the town. This time, emphasizing the bordello."

"Sure, I have no objections."

Pam attempted to curb her enthusiasm while she studied her partner. Finally, she recognized his tired eyes and the slump of his shoulders. "You look beat, Nolen. Have you rested lately?"

"About four hours a day, since we filed in Susanville. Getting the mining crew started, orchestrating the special

Board meeting to oust Roger Benjamin, and extracting the first load of ore has kept me on the run. The last thing I must do is talk with Paul. Something about what's happening in the town. Then, I'm going to crash."

The pilot raised his voice, to be heard above the rumble of the helicopter engine. "Sir, after we rise above the approaching hilltop, you'll have a splendid view of the valley. I think you'll find the scene quite interesting."

Nolen and Pam looked forward through the plastic nose bubble. When the aircraft shot over the crest, the two passengers appreciated the pilot's suggestion. The high desert, usually still and dark at night, had been transformed. Hundreds of red and gold flickering campfires dotted the mountain hillsides. The fires marked the locations of the many gold hunters, already scouring the land. In the center of the flat valley, lines of automobile headlights illuminated the roads leading to the boomtown.

"It sparkles like a brilliant pendant," Nolen whispered.

Pam laughed, then said, "My crew will film this picture later tonight."

I wish, Nolen thought, that I could enjoy this sight with Maida. Now that the first shipment departed, I can focus on her. I'll get some sleep first, so I can think straight. Then I'll find her, and hopefully, rebuild our relationship.

As they approached the town, the radiant scene decayed. A harsh reality became apparent, when the helicopter descended into the hemisphere of light and circled over the compound.

The boomtown was separated into quadrants by two intersecting four-lane streets. Cross Street bisected the compound from east to west. Stateline Avenue traversed the California-Nevada border from north to south.

Cars and trucks were allowed into the town only at the east and west ends of Cross Street. At those entrance gates, people wishing to remain in town were directed south into the parking lots. Prospectors seeking supplies, tools, or fuel moved forward into lanes passing in front of the gas pumps and warehouses. All vehicles leaving the boomtown exited north along Stateline Avenue. By Nolen's design, a one-way stream of traffic flowed into the compound, met at the main intersection, and then departed.

To control access into and within the boomtown, fences circled the town, the bordello complex, and each of the storage lots behind the two warehouses. Only foot traffic was allowed among the buildings south of Cross Street, where swarms of people moved like rippling waves.

Pam pointed to the center of the town. "Look. Between the bars."

A crowd had formed a circle in the street, surrounding two men who were slugging each other. As the aerial observers watched, a third man stepped forward to assist one of the fighters. Quickly, a brawl erupted among thirty men. Moments later, three security jeeps, with blue and yellow lights flashing, raced south from the headquarters building, and closed in upon the fight scene.

"What a carnival of life," Pam exclaimed.

A distinct change overtook Nolen's demeanor, as he viewed the commotion. "Pilot," Nolen ordered, "I want to walk through the town. Land at the southern end of the compound, inside the fence, within the RV parking area."

Pam protested. "Nolen, it would be better for us to go directly to the headquarters building. Paul is waiting for us at the conference room. And you need to rest."

"Pam, until now, I've been motivated by Digger's dreams, Hilda's hopes, and my plans. All of those things were theory. I'll rest after I see what we have created, after I feel the reality of the town."

As the aircraft settled onto the ground, it kicked sand onto the vehicles and people in the parking area. After its departure, Nolen and Pam heard men cursing them, for landing inside the fenced town. Ignoring their angry shouts, they walked toward the base of Stateline Avenue.

The two looked north, up the half-mile overcrowded avenue, to study the milling throng. They saw men, mostly. Periodically, smaller, strong-willed women shouldered their way through the clamoring crowd. Lines of impatient consumers snaked along the front of every building, anxious to enter, to buy liquor or food or a hot shower. Knots of gold seekers crisscrossed the center of the street, laughing, shouting, cursing. The many men and women created a rising and falling rumble of sound.

Nolen and Pam bumped along the street inside a kaleidoscope carpet of humanity. Tall people, short people, fat people flowed in all directions around them. Some were happy. Others bellowed anger. Most were clothed in heavy-soled hiking boots, dirty denim jeans, wool shirts, down jackets, and some type of hat. Some, wearing lightweight coats and city shoes, reflected their recent arrival to the gold fields, as well as their ineptitude to prepare for the rigors of prospecting.

Nolen watched a town worker, uniformed in orange overalls, stop a mis-dressed individual and explain that appropriate outfitting could be purchased at the warehouses, located along the northern end of town. The workers were doing their job, ensuring no opportunity was overlooked, to sell all that the boomtown had to offer.

The blanket of dust covering the jostling crowd was a surprise to Nolen. Sprinkler trucks, which wetted each street hourly, were no match for the thousands of shuffling feet. Nor had he expected to see the quantity of discarded paper, aluminum cans, spoiled food, and cigarette butts littering the ground. Nolen noticed, Paul had added refuse teams, with rakes and plastic bags, to keep ahead of the crowd's mess, before it became a sanitary problem.

"It's amazing what we've unleashed," Nolen remarked. "Seeing it first hand, sends a thrill through me."

"Smelling it, turns me off," Pam grumbled. "Still there's a multitude of destinies around us. And I have been tapping into them before they disappear. Each morning and afternoon, I have been filming human interest interviews with people from varied walks of life who have raced to the boomtown. Sometimes, I ask them to explain how they plan to find gold. Other times, I encourage them to elaborate on why they left their jobs and families, when the chance of finding gold is so small."

"You sound like a kid in a candy store," Nolen commented.

"Yes, it's a news reporter's dream." Pam chuckled, then continued explaining. "One woman remarked on camera that she would rather take her chances here, than stay with her jerk of a husband, back in Sacramento. Then, there was a couple whose car was shoved into a ditch by angry prospectors, simply because it had stalled in an intersection. I even filmed a teenager who found a nugget in a mountain stream. For such a small discovery she had great dreams, rather like yours, Nolen."

"How have those stories been received by the TV audience?"

"Stations throughout the country report the public can't get enough of my stories. Probably, because TV allows people to vicariously participate in the gold rush. Especially, those who can't come here, or those who lack the courage to take a chance. Because the interviews are so hot, my station manager suggested I locate two unrelated persons, from the Bay area, headed for the boomtown. I'm now delivering a noon-time segment about this man and this woman. The first episode caught them leaving the city and traveling north. Soon, they will be prospecting, and deciding whether to stay or leave. It would help entice more people to the boomtown if one of them hit it big!"

"Keep me informed on their progress. Possibly their destinies will require a little assistance from our Association."

Boodan leaned against the pickup. He carefully sighted through his rifle telescope, aligning the cross hairs on the passenger door of the helicopter, as it skimmed along the airstrip.

At his shoulder, Luke studied the aircraft with field binoculars, while it settled to the ground 200 yards away, in front of the hanger at the northwest edge of the boomtown. Both manhunters watched the pilot and copilot leave the parked helicopter, then stroll into the hanger. "What happened to Martin?" Luke growled.

"Shit!" Boodan swore. "I thought we finally had him after spotting his ugly face, peeking through the helicopter window, when they flew by toward the far end of town. They must have landed and let the bastard out."

"So, what do we do now?" Luke sarcastically whined. "Your brilliant idea that he would come to town, and be an easy target when he flew into the airstrip, was only half brilliant. That SOB has slipped away again. I'm tired of freezing my ass off, waiting in this dirty gully to ambush him."

Boodan adjusted his red eye patch, as he considered alternatives. "Well, he's on foot and must be headed toward the headquarters building. That means, we know the last portion of the route he has to follow. It will be just like sittin' in a deer stand."

"You're right! He'll be forced to walk north, along Stateline Street. He has to enter the only personnel gate into the fenced storage yard. That channels him into the open, when he walks from the gate to the front door of the headquarters."

Boodan nodded his head in the affirmative. "We'll have a clear view of the path he will follow. Then, POW! We knock him down. And if any security guards are around him, we hit them too. That should cause enough confusion, so we can run back here and escape to the west. But we gotta' get into position before he reaches the storage yard. Quick! Get your gear. We'll cut through the storage fence and position ourselves for an easy shot."

Luke reached into the cab, seeking a silencer for his rifle. Boodan searched for the wire cutter in the tool box of the truck. "Martin will be dead within an hour," he said. Then the Cajun began laughing.

38

Goldtown

30 September

Nolen and Pam studied the garish, five-building bordello complex. Each doorway, window, and roof were outlined with a string of alternating red, white, and blue twinkling lights. A trio of flashing neon billboards, spaced almost sixty feet apart, enticed town visitors to enjoy the pleasures available inside. Below each billboard, a swath of red lights encircled one-way entry gates, which controlled the flow of men into the area.

Those entering the gate were scrutinized by security personnel, who used hand held metal detectors to confiscate knives and pistols. Seized weapons were tagged, bagged, and sent to the exit gate at the southern end of the section. There, they would be returned to the customers when they departed. After passing through the weapon inspection station, each customer purchased tickets for the theater or for one of the brothel houses.

A serpentine line of several hundred men stood along the fence, waiting to pass through one of the entry gates. It was obvious that many of the boisterous, laughing males had been boozing for some hours. Several men noticed a group of scantily clad women, lingering near the side door of the brothel theater. The men hollered crude descriptions of the good times they would soon provide the whores. As Nolen and Pam passed the men, a few saw the female reporter and self-consciously diverted their gaze. Those who were bolder openly leered at her.

Pam pointed toward the front of the line. "What's that tall guy doing?"

A thin man, dressed in black, walked along the line of men, shaking hands with some, saying a few quiet words to others. Suddenly, two young men began shouting at the older man.

"Seems like the boys are becoming a bit rowdy," Pam commented.

The two youths shoved the tall man to the ground. He lay quiet for a moment, then reached out his hand to his assailants. They responded by kicking him. Their heavy boots were harsh. He winced with pain. As the fallen man rolled away from the bitter blows, Nolen glimpsed his white cleric collar. "Let's help this guy, before he gets killed."

When Nolen reached the man, he clutched the back of his coat and pulled him to his feet. Nolen yelled over his shoulder, "I'll get rid of the nuisance." He then dragged the protesting man from the gateway. After moving some distance from the bordello line, Nolen stopped. "You don't fear taking risks, do you, Padre?"

"My name is Father Gabe Jones." The priest brushed dirt from his clothing. "Risks are part of my vocation. But God watches over me. And apparently so do you. Thanks for helping."

"Why were you at the bordello gate?" Pam asked.

"When this boomtown sprang up within my Catholic parish, I asked myself, why here? I'm the only priest within fifty miles. Why me, when I was relegated to this deserted region by a Bishop who doesn't trust me? He considers me a fake, because I came to the priesthood late in life, after being a great sinner. He questions my ability to be a good shepherd."

"So, what are the answers?" Pam questioned.

"God's given me the task of comforting the weary souls, who are racing here for wealth, overcome by greed. God knows I can empathize with these people flooding the valley. Years ago, I chased carnal and material things. Yet, I was always unhappy, yearning for peace. Regrettably, before I found my peace, I contributed to my wife's drug overdose. I believe God wants me to be his messenger now—to tell the story of a greater salvation, more precious than any man's gold."

"How well have you been delivering that message?" Nolen inquired.

"I do try. Each night, I stroll the streets of the town, introducing myself, offering to listen to anyone who needs to talk. I explain how I used to be just like them, but I discovered a better way. Friends, I've learned that bordello gate is the perfect source for finding intolerant sinners, and lost souls, aching to be saved."

"Is Gabe short for Gabriel?" Nolen wondered.

"Yes, I took that name when I was re-born seven years ago. It gave me comfort, while I was the oldest novice in the seminary. And now, ironically, it seems to fit me well. For here I am, patrolling the sinner's gate, blowing the Lord's horn. A Gabriel, attempting to call my brothers back into the comfort of God's church."

"Father Gabe," Nolen said, "if you continue to walk the bordello lines, you could get seriously injured. You'll incite fights between men who don't accept what you represent, and others like myself, who don't enjoy seeing an old priest shoved around. So, if I arrange for a large tent to be erected in one of the open areas inside the town, would you promise to stay away from the bordello? The tent would serve as a church, where you could hold services, discussions, or whatever you like."

"Son, I'd say the cost of one square foot of property inside the boundary fence is exorbitant, far beyond my pocketbook of poverty. How do you come by the power to offer me such a site?"

"I'm Nolen Martin, majority owner of this town, as well as the gold vein that caused it to be developed. We were reserving the open areas to build additional bars or flop houses, once we earned enough money from the existing businesses. Letting you establish a church tent will bring some good into the town, and perhaps, dampen some of the chaos."

The tall priest barely smiled, as he filtered his fingers through his sparse white hair. "Does this motivation come from true kindness? Or does it come from a mercenary desire, to prevent your bordello profits from being invaded by a lunatic priest, ranting and raving beside those red gates to Hell?"

"Father Gabe, you will have the sole church in the boomtown. You can spread your gospel, while retaining the generous donations from the prospectors, at each service you conduct. Once more strikes are found, you will be collecting a wealth of money for your church. Possibly, even become my confessor. Wouldn't that please your Bishop?"

"Son, at the seminary, I learned good and evil come in many disguises. Devil or angel, which are you?"

"Every man is a combination of both," Nolen answered. "And every man must decide which one will dominate him. Consider my offer. Decide whether I'm aiding your cause, or bribing you to look the other way. Then, visit me later tonight at my headquarters. You can tell me whether you accept or reject this opportunity."

Nolen and Pam left the priest to continue walking north, between the bath house and the flop house, bumping shoulders with the many individuals pushing past them. A gangly, fifteen-year-old black youth, waving a one-page newspaper, fell into step beside Pam. "You folks want a copy of the *Nugget*? Just one dollar for all the latest news, 'bout what's happenin' during the gold rush. Each side of the page, gots lots of info. It tells ya' how to find gold, the prices for supplies, and how to avoid them con-men."

"I could have used such information weeks ago," Nolen chuckled.

"For another dollar, you gets a map showin' the best routes 'round the jammed roads in the county. On the front page, there's the latest claim sites, recorded at the county seat. On the back, there's an obit...obituarry. Tells who's been killed in the hills. This paper's a bargain, 'cause it'll save ya' huntin' time. Maybe your life, by keepin' ya' from puttin' down stakes on somebody else's claim. How many copies you wants?"

Pam shoved two dollars into the boy's hand. "Who publishes this paper?"

"Lady, how much ya' payin' for what I knows?"

Pam waved another two dollars in front of the youth.

Smiling, the boy answered. "I gets my papers down the street on the right, just before you reach the gate. All the newsboys buys them from the man in the tent. If we sell 300 sheets a day, he gets us a free meal. I likes this great job. Better than livin' on the streets in Oakland. Lady, does your friend need a paper?"

"No thanks kid," Nolen said, as he took the sheet from Pam. "Hum, *Nugget*. That's an appropriate name for our town paper."

"This guy is clever," Pam commented, scanning the length of the five-lane avenue. "He's hired a bunch of homeless kids to distribute his paper. Look, the youngsters are selling to drivers in cars, creeping past the gas pumps and warehouses. They are also soliciting those leaving the parking lots and departing for the gold fields. That gives him two ideal opportunities to sell to everyone passing through this town."

Nolen stepped in the direction of the tent. "Let's meet this innovator."

"Damn," Pam grumbled. "He must be raking in a fortune. It costs him pennies to produce and distribute each paper, that people are willing to purchase for one dollar. He's bringing in three, four—maybe even $5,000 per day. That's $90,000 to a $150,000 a month. Wow!"

"Don't begrudge him his profit, Pam. I'm sure Paul made this guy sign some contract, to ensure we get a piece of the action."

"That's not what's bothering me. I'm upset because I didn't think of this idea myself."

A hastily-built wood fence surrounded three tan-colored tents, keeping passers-by away. At a ticket stand, a barker chanted. "Step into the center tent, folks. Have your photo taken. Just seven dollars, to record for history, that you actually were part of the California modern-day gold rush.

"We have spared no expense to keep you informed. Behind the tent to my right, we have erected a satellite dish receiver. Don't miss the opportunity to appreciate the latest national news. For a mere four dollars, you can learn what the world is reporting about the gold rush.

"Moreover my friends, if you only want to read about the activity in the surrounding gold fields, move down to the tent on my left. Buy your copy of the local gold field tabloid, the *Nugget*. For the paltry price of one dollar, you will procure the latest news of gold strikes, fights, and sights.

"Or, save three dollars by purchasing our super ticket, allowing admission to all three tents. Have your photo taken, see the latest world news, and get a copy of the *Nugget*. All for one small price. Come on folks, step right up!"

"Hey, pal," Nolen shouted to the barker. "Where can I find the owner?"

"Buddy, I'm busy selling tickets. I'm not my boss's keeper. Besides, what's it worth to you?"

Pam dangled her news credentials in front of the man. "It's worth your job, if he learns you didn't help him receive some free publicity. I'm with TV Channel 25. My purpose here is to coordinate a feature story about this business."

The man frowned, hesitating, then said. "Go to the rear of the newspaper tent. Tell the man at the door, that the barker sent you to see Solomon Kohl."

Young Solomon Kohl, and two other men, were kneeling between two printing machines, studying the gears of one, when the doorman interrupted their concentration. Upon seeing the couple standing in the tent doorway, Solomon quickly wiped sticky grease from his hands and hurried over.

"You're Pam Pendleton. Everyone in America watches your great newscasts. You look very attractive on the screen. How can I help you?"

Pam eyed the youth's curly hair and aquiline features, enjoying his compliment. "My partner, Nolen Martin, wanted to meet you."

Nolen shook hands, "I didn't expect someone in their twenties."

"Nineteen, but don't let my age fool you. I've been buying, selling, and hustling, ever since I began working in my dad's jewelry shop. I was just thirteen. Since I've been in college, though, I don't say hustle anymore. Now, I'm an entrepreneur. I've learned that phrase makes it easier to raise capital for risky ventures."

"I'm curious. How did you come up with this scheme?" Nolen asked.

"A few days ago, I was walking out of my business class at Fremont College. A friend told me he had just heard about the gold strike. When I saw Pam's TV shots covering the jammed streets in the boomtown, I recognized the niche market for a local news service. I decided it would be far more lucrative to get into the action here, rather than sit around listening to some boring professor teach marketing theory. I was right."

"You established this print shop quickly," Pam commented.

"Who knows how long this gold rush will last? I had enough chutzpah to yank my father's satellite dish from his back yard. Then, I was successful in persuading an uncle, who owns a small printing service, and a cousin, who earned his living photographing bar mitzvahs, into joining me. Using my uncle's flatbed truck, we delivered all our tents and equipment here in sixteen hours."

"Your professor would give you an A for initiative, and an A+ for creativity," Pam remarked.

"Never have I worked so hard in my life. My first problem, after arriving, was finding Paul Martin. Eventually, I did. We negotiated an exclusive lease. I only have to produce $120,000 worth of sales during the first month of operations. So, here you see it. The beginning of the Kohl empire. In full operation. Or, at least, it will be, when my uncle gets that collating mechanism working on our oldest press."

"Other than my discovery," Nolen asked, "have there been any other strikes reported during the past three days?"

"Of the 262 claims already submitted at the county courthouse, my field reporters tell me that fourteen claims were based on some gold being found. Enough, in one case, to cause a gun fight and the mysterious disappearance of three people. By the way, most of those positive sites are uphill from your gold vein."

"Please instruct your news gatherers to report the size of each confirmed strike," Nolen stated. "Each Friday, supply me with a weekly summary, designating the exact locations where any gold was found. I am particularly interested in what Continental Mining and Refining is doing."

"CMR was the first company to file after you recorded your claim. Their four sites surround your discovery. If they hit a valuable vein, paralleling yours, you two will be doing business side-by-side for years to come."

"I was afraid that would happen," Nolen remarked.

"Mr. Martin, in exchange for my assistance, perhaps you'll give me some help. Tell me, what will you name your mine? During the first California gold rush, most mines had colorful names like 'Prosperity' or 'Glory Hole'. I could use an exclusive scoop to spice up tomorrow's feature story."

"Good point, Kohl. I've been so busy, I haven't given any thought to naming the mine or the town. Unique titles will give us recognition, especially when news articles or stocks are released."

Pam slipped her arm around Nolen's waist. "A woman's name would sound good, Sugar."

Nolen nonchalantly slid away from Pam's embrace by picking up a copy of the town newspaper, setting on a nearby table. After a few seconds contemplation, he answered, "For years, my foster father maintained his dream of finding a huge strike. My partners also embraced the vision of success that I kept dangling in front of them. Now, thousands of people have come here believing they can quickly alter their future. So, we'll call the mine 'Digger's Dream'. And 'Goldtown' is an appropriate name for our boomtown."

"Thank you, Mr. Martin. When I deliver the weekly summary report, possibly you can again share some colorful ideas with me, concerning your business activities. That will afford something of interest to satisfy the prospectors who fill your town."

"That depends," Nolen smiled, turning toward the doorway, "on how good your information is."

Nolen and Pam proceeded north, past the Nevada gate. There, extended rows of automobiles and trucks inched by the fuel pumps, warehouses, and stores. Although it was late at night, an efficient assembly line of men and machines serviced the string of vehicles. Orange forklifts, hauling pallets stacked high with stock, rumbled out of the buildings onto the issue docks to deposit their loads. Scurrying clerks hustled along the docks in a clockwise pattern, filling tub-sized, red plastic carts from the pallets. Periodically, they reviewed a customer order slip before hastily loading a pick or a shovel or another item listed on the form.

Nolen studied the hopeful desire in the eyes of the many men and women. Eyes not yet dimmed with fatigue, hunger, and pain, or worse yet—tainted with despair and failure.

He wondered who among the throng would become the great winners, with the good fortune to hit a large run of gold during the coming year. Nolen knew that almost all would fall short of their bloated expectations. Within a few short weeks, the weaker ones would be winnowed down by the reality of back-breaking, hand-blistering work, expended on meager claims. The cold mountain elements would relentlessly lash them, while the enticing warmth of hearth and home would beckon but a few hundred miles away.

So easy to climb into their vehicles to drive home, forever avoiding such self-inflicted drudgery. Just return home to friends and family, and pay the price of re-admittance into their former routines. The price would be explaining why they had failed, or providing a plausible tale about an injury or illness, sustained while in the hills. Some of the men and women uprooted by the gold rush would be unable to return and face their past. Instead, they would drift on to new destinies.

"Pam, I've seen enough of Goldtown," Nolen declared. "Let's have one of the security teams drive us to headquarters."

The middle lane of Cross Street was reserved for security traffic. As their commandeered jeep bounced along, Nolen mulled over the activity. Every few seconds, a prospector rolled his vehicle away from the last supply building. Then the prospector would either turn north onto the exit road, or turn south into one of the two parking lots. The jeep driver swung north between the banks and then turned left, passing into the California storage area. He sped away from the bright lights of Goldtown, toward the Headquarters Building, to deliver his important passengers.

A scarce hundred yards west of the headquarters entrance, Luke lurked in the shadows, behind a stack of lumber. He restlessly stamped his feet and rubbed his arms to keep warm. Several feet away, Boodan was pissing against a barrel. "Vehicle coming," Luke said, unconcerned, watching the security jeep pull through the compound gate.

"Use the binoculars, damn it," Boodan ordered.

"Looks like another load of guards. Can't see any faces yet."

Boodan zipped his pants, and stepped toward the rifle stand.

Luke swore. "Damn! He just hopped out of the jeep."

Both men snatched up their rifles, wrapping their arms into the rifle slings. They dropped their elbows onto the lumber pile and squinted through their telescopes, desperate to view Nolen among the moving figures on the far side of the jeep.

"Shit! The jeep is blocking my shot," Luke growled as he sidestepped, seeking a more favorable opportunity.

"Don't fire unless you have a clear shot. Be certain you can kill him with your first round," Boodan declared. "If we miss, he'll go into hidin'. Then, we'll lose our chance of gettin' him."

"The front door is open! Can you get a bead on him?"

"Maybe," Boodan whispered. "Just...hesitate...one second. You bastard! Shit! He stepped through the doorway, before I could squeeze!"

"Well, what the hell do we do now?" Luke snapped.

Boodan laid his rifle against the lumber pile. He took a deep breath, to let the adrenaline seep from his body. "We ambush him when he comes out later tonight. Or, if we have to, early in the morning, when everyone is asleep, we sneak in there and knife him.

"For now, move off to my left. Go about 100 feet or so, where you can cover the rear exit. I don't want him slipping away again. If he walks outside, use your radio to alert me, then shoot him if you can. Pump at least three bullets into him, to make sure he dies. I'll do the same thing, if he appears in this direction."

39

Present

Lights flashed on within the conference room, as confetti showered Nolen. The room was stuffed with revelers, dressed in party hats, shouting, "Surprise!" Balloons and streamers decorated the ceiling. A table in the center of the room held a white and gold, three-tiered cake, together with a fountain, bubbling sweet smelling champagne.

Nolen laughed and turned to Pam. "Were you aware of this?"

"Of course," she replied. "I was supposed to have you here an hour ago."

"Enjoy it!" Paul shouted. "The difficult work is over. All we must do now is relax, and count the dollars rolling into the bank. We deserve to celebrate." Paul waved his hand at the packed room. "So, I gathered together most of the Association partners and their families, plus key politicians, and our friends. Kemper and Jake are the exceptions. Kemper is still at the mine. Jake had to handle some disturbance."

"Yeah, I saw the ruckus just before we landed." Nolen quickly scanned the crowd for his precious Maida, then hid his aching disappointment.

The men and women crowded close to Nolen, shaking his hand, slapping his back, or kissing his cheek. All were grateful for their share of the windfall of wealth flowing from the gold rush.

Paul called for attention. "Quiet everyone. I think it's time to tell my little brother what has transpired, while he's been lazing around the mine. Most importantly, the town is not nearly large enough to handle the demand, so we doubled our prices today."

The group cheered and clapped over the news.

"Paul, I told you, you would enjoy this job," Nolen stated, as the cheering subsided.

"Moreover," his brother continued, "we've made several other improvements in town operations. To promote the gamblers' length of playing time, I followed Pam's suggestion to move some prostitutes into the casino. Additionally, tomorrow, I commence building permanent residences north of town. Soon, each Association partner will regain some privacy in their lives."

Chatter and suppressed laughter swept through the throng, as they digested the impact of the improvements.

Hamilton eagerly interrupted. "Half the lots adjacent to our town have already sold. I've included a franchise clause in each contract. Anyone establishing a business within the town limits must use our banks, as well as pay us a percentage of their sales. Every hour, tents housing the new businesses are being erected. The town will triple in size during this next week."

"Amazing," Nolen replied. "The gold rush is far more expansive than I ever expected."

"The best news," the banker continued, "is our excellent cash flow. All our debts will be completely satisfied within twenty-five days!" The crowd again cheered. Senator Santreno began an impromptu jig.

"Furthermore, friends, you're looking at the new President of First Mountain Bank. Yesterday, I pressured the other bank directors with the withdrawal of all the Gold Association accounts, unless I were appointed President."

Nolen hugged his partner. "Your wife should be pleased with your accomplishments," he whispered into Hamilton's ear.

Pam addressed the group. "I also have positive news. Earlier today, our Los Angeles lawyer informed me that our movie and book rights have sold for five million dollars—plus a percentage of sales!"

It took Paul several minutes to somewhat quiet the rambunctious congregation. "Many of our suppliers are attempting to void their obligations," Paul finally explained. "The companies from whom we buy food, liquor, and materials have been crying for more money. I warned them we would enforce the penalty clauses in their contracts, if they failed to deliver. When they signed those supply agreements, prior to the gold rush, they didn't think they would ever want to renege on such lucrative deals. As a consequence, they accepted the failure-to-deliver penalties which we included. I've been reminding the reluctant firms, that they will lose all our future business unless they fulfill their contracts. They believed me when I told them how many other merchants were willing to support us. Consequently, we've had few supply delays."

"Everything I've heard tonight is splendid news," Nolen said. "I have some more. As we speak, the first load of gold is en route to the smelter." Everyone in the room responded with overwhelming applause, whistles, and laughter.

Waving her arms, Hilda attracted the group's attention. When their exuberance subsided, she spoke. "Nolen, we appreciate everything you've done. Your determination drove all of us to not hesitate. You encouraged us to achieve our dreams. As a token of our gratitude, we have a small present for you."

Part of the crowd squeezed away from a side door, opened by a waiter. Maida entered pushing a cart. The cart transported a large box, decorated with red wrapping paper and topped with a gold bow. Two holes were cut into the side of the container, which jiggled as something moved about inside.

Tears glistened in Nolen's eyes when he saw Maida. She was as radiant as ever. He finally felt a glimmer of hope that he would win her back.

"Open it," Maida urged.

Nolen released the bow and slid the top away. A floppy-eared, black retriever puppy placed his fuzzy forepaws on the edge of the cardboard box. He uttered a high pitched yelp and panted, as he watched all the guests staring back at him. Nolen cuddled the small animal in his muscular arms. His voice cracked as he spoke. "Thank you. Thank you all. This means a great deal to me."

The phrase, Someone loves me, surfaced in Nolen's mind causing him to smile with satisfaction. As the group applauded, he let the coldness within him begin to fade.

"Son, it's time to relax," Hilda said. "You have fulfilled the promise you made. Once the doctors were informed about a possible biological agent, they modified Digger's treatments. Today, he opened his eyes for the very first time. I told him all the great news. He's so very proud of you. And of course, I am too. Take some time off. Let the Association work for you, now."

Nolen hugged his mother. "Hilda, I'm so happy to hear that Digger is recovering. Tomorrow I'll locate the best cancer doctor in the country, and fly you to him this week. Everything will be better for Digger, and for you."

Maida waved her bandaged hand at her noisy partners. "The value of our mining stock keeps rising. Tomorrow, when we announce the results of the first smelting, it should hit $100 a share. That's sixteen times higher than the six dollars a share we originally paid. Furthermore, earlier this afternoon, the spot price of gold quit dropping. It's down eighty dollars. As I predicted, I'm now receiving calls from investors wanting to get out from under their commodity contracts."

Nolen set the small dog down and pulled Maida into his arms. He kissed her. Their friends whistled and clapped, encouraging their continued intimate embrace. Paul signaled the DJ to begin playing dance music. The celebration commenced. Waiters circulated, offering slivers of cake and crystal glasses filled with champagne.

Nolen whispered, "Maida, Maida, Maida. It's so wonderful to see you again and to know you are well. I was afraid I had lost you!"

Maida took Nolen's hand, and led him away from the center of activity, so they could visit privately. "I'm sorry you thought we were through, Nolen. But it hasn't been easy for me." A brief shudder shook Maida. "I had to conquer the helpless feelings and overcome the trauma of being kidnapped. I needed time to think about what is important to me."

"What did you decide?" Nolen cautiously asked.

"That you are an essential part of my reason for living. That you are more precious than the difficulties, or the danger."

Nolen laughed and clasped her. "I promise—you will never be hurt again."

Maida's seductive smile returned. She kissed Nolen. "When do you and I start spending more time together?"

"Beginning tonight, Precious. This weekend, we'll take our first vacation. Where would you like to go?"

"Someplace I've never been. A place with swaying palms and a sandy beach. A hideaway where no telephones or reporters can bother us. But that may be difficult to find, now that you are the most coveted celebrity in America. It will be an effort to retain any privacy. Women will flock to you, attracted by your money, the allure of excitement, and..."

"And my exceptional good looks?" Nolen interjected.

"Yes, that too."

"Concerned over the competition?"

"I want our relationship to flourish, not flounder."

"Don't worry, pretty lady. You have so much more to offer than other women."

Maida impishly smiled. "Name six things."

"Most of all, you cared for me, before I became rich and famous. You teamed with me to achieve a common goal. You also trusted me."

Maida playfully encouraged more compliments. "Is that all?"

"You have a warmth which tempers my cold side. And I admire your courage. I'll never forget how you protected the Association's secrets when you were kidnapped."

Maida stroked her hair and giggled. "It is satisfying to learn you finally recognize a few of my unique qualities."

Nolen baited Maida. "Well, sometimes you are fair in bed."

"Fair!" she whispered. "I have to muffle your moans with a pillow each time we make love, just so the whole neighborhood doesn't become alarmed."

Nolen laughed. "Could I ask you to do one more job for me? It's safe, and it's easy."

"Sure, Lover, what is it?"

"Establish a scholarship fund for poor Latino youths in Los Angeles. Transfer one percent of my boomtown profits to that fund for the next five years. Create a board of directors. Ensure that our friend Cholo Cantera becomes permanent Chairman, with a handsome annual salary. Also, assign him veto power over all policy decisions and financial grants."

"I'll complete the task before we leave on vacation."

Nolen handed a key to Maida. "Please, call Cholo tomorrow. Tell him there is a sealed envelope in this safe deposit box. The envelope contains the pathology data. He can use it any way he sees fit. Tell him, within six months I'll destroy the only other existing copy. That is, if the police and government do not disrupt our mining, stock, or Goldtown efforts."

"Certainly. I'll speak with him tomorrow."

The puppy seated at the miner's feet whined, causing Maida to stoop down. "It sounds as if this little fellow may need to go to the bathroom. Let's take him out back, to the storage yard."

In the ebony shadows at the edge of the headquarters building, Luke shuffled back and forth, waiting. He heard the rear door to the headquarters squeak as it opened. Maida stepped into the storage yard and set the puppy free. The energetic pup scampered toward a nearby lumber pile, where he began to sniff. A startled rat popped from behind the woodpile. Suddenly, both animals raced further into the darkness. Nolen followed, through the doorway, running after Maida and his new pet.

Luke grabbed his rifle, but he was unable to aim before Nolen also disappeared into the black storage yard. Luke urgently spoke into his hand held radio. "Boodan, Martin's in the storage yard with that bitch we kidnapped. I didn't have a chance to fire a shot. He and that woman are chasing a dog."

"What?"

"I said, a dog, a puppy."

"Wait for me. I'll be right there. This is it, our best chance to kill both of them."

Piles of tin roofing, lumber, tentage, dusty cement bags, ten-gallon water cans, wooden pallets, and numerous other supplies were stacked throughout the storage area. The uneven rows of material filling the yard varied in height. In the middle of each alley, light from a full moon aided Nolen and Maida to see where to step. Near the alley edges, dark shadows hindered their ability to identify where the puppy had crawled, in urgent pursuit of the elusive rat. They hesitated, listening for any sound from the young dog.

Nolen became aware of the steady, noisy murmur created by the activity in the main section of town. He felt uneasy, automatically recalling the command which had protected him numerous times during the previous, hectic and dangerous weeks.

Fight to Survive!

Deliberately, he ignored his instinctive mental alarm. I've succeeded, he thought, rationalizing away his apprehension. Everything has progressed as planned. My conditioned response is no longer needed.

A timid growl from the right seized the couple's attention. Maida noticed a narrow opening between precarious stacks of filled water cans. Only a few white shafts of light trickled into the tight passage through which they squeezed.

"Come here, puppy. Come here," Maida softly called. They heard the pup's short, high-pitched yip followed by the sound of paws, rapidly scratching. Maida soon identified the outline of the small animal. His nose was poked between two wooden slats. She knelt, and began petting the pup.

Nolen heard a voice behind him quietly grumble. "Where the hell did he go? I thought I saw him when we stepped into this alley."

Nolen looked over his shoulder, spotting two men whispering in the moonlight. A chill rippled through him upon seeing the stalkers. Suddenly, he recognized the attacker he had subdued at the gas station, and the shorter man he had knocked unconscious at the house in Sparks. The bigger one moved. Moonlight reflected off gun metal. Nolen's buttocks tightened.

Think and stay calm, his instincts dictated.

Nolen did not disregard the conditioned reflex this time. Instead, he held his breath, hoping he and Maida wouldn't be detected.

"Ruff, ruff, ruff." The puppy barked in Maida's arms.

"He must be in there, Luke!"

Nolen rammed his shoulder against a stack of water cans, shoving with all the strength that his muscled legs could deliver. Ten-gallon canisters cascaded toward his assailants. Nolen spun, as he pushed Maida through the end of the tight passageway, struggling up and over several cans. The couple spilled into another alley. The snapping crack of bullets, breaking the sound barrier above their heads, in unison with exploding water cans, forced them to remain on their hands and knees.

"Follow him, Luke!" Boodan yelled. "I'll cut him off."

"What are we going to do?" Maida gasped, frightened.

Nolen dragged Maida around a corner. Quickly, he pushed her into the shadows between two stacks of pallets. "I promised you, you wouldn't be hurt again. Hide here, while I divert their attention. Then, run to the headquarters, to get help. Tell Jake they have silencers on their rifles. Remember, I love you, Maida."

Nolen raced toward a six-foot high pile of lumber. He leaped up the stack of wood and grabbed the top edge. Pulling with his shoulders, he swung to the top, ignoring several exposed, rusty nails which ripped open his right palm. Any second, he knew, the man hounding him might be able to see well enough to shoot again.

Suddenly, Nolen spied the shadowy figure chasing him. He yelled, "I'm over here, you big ape!" Two bullets seared the air near his head, as he rolled from the far side of the lumber pile.

Nolen crashed onto the ground. He was exposed in full moonlight, vulnerable in an open section of the yard. Nolen recognized that if he attempted to cross the compound and climb the wire fence, he would be clearly visible, an easy target. Quickly, he searched his immediate surroundings for a weapon. Several broken pieces of half-inch, steel reinforcing rod lay near his feet. He grabbed a four-foot long shaft, and crouched with his back against the stack of wood.

A cold trickle of sweat wiggled down Nolen's spine. His fear mounted, as he listened to his pursuer struggle to the top of the lumber pile. Luke swore when his hand also grasped the exposed nails. Then the CMR man stopped moving. Only his labored breathing betrayed his proximity.

He must be scanning the area, Nolen thought. Just then, a boot toe poked over the edge of the pile. Nolen reacted, jumping to his feet and ramming the jagged rod upward, into Luke's exposed crotch. "That's for killing Pal."

A gut-tightening scream ripped the air, followed by the muffled discharge of the rifle in Luke's hands, when he involuntarily tightened his finger on the trigger.

Nolen slammed the rod a second time into Luke's crotch. "And, that's for mauling Maida!"

Staggering backward, Luke clutched for the severed artery in his leg before tumbling off the far side of the lumber stack.

Nolen had no inkling which direction the smaller man had taken. Time was evaporating. Hoping he was racing away from the killer, Nolen turned right, and rushed toward one end of the line of material.

He was wrong. The Cajun stepped into the open, twenty feet in front of the miner. Nolen slid to a halt. His heart pounded, as he stared at the deadly, silencer-tipped rifle casually held in his enemy's hands.

"Well, well, well," Boodan sneered. "We meet again. Remember me? You gassed me at the duplex. Now, it's my turn. I'm goin' to blow your shit away!"

"Don't shoot. There's nothing you can gain by killing me."

"Sure there is, kid. Durin' the last few busy weeks, you overlooked something real important. You made a big mistake by forgettin' your personal affairs. You haven't changed your will."

"I don't understand. What do you mean?"

"You're still married. Everything you own, your wife will inherit, after you die. And wouldn't you know it, CMR made a deal with her. We get half the estate when you croak." Boodan laughed with wicked pleasure.

"Listen to me! Your partner's screams have alerted my security teams. They will arrive any moment. You can't get away with this."

"Bullshit! The noise from the town muffled Luke's cries. Anyways, they won't get here in time to stop me. This is one killin' I'm gonna' enjoy."

Nolen watched the rifle swing toward his heart. He leaped sideways. The experienced manhunter had anticipated just such a move. Nolen did not hear the muted blast. He saw only the blinding muzzle flash, and felt a hot, sledgehammer-blow slam into his left shoulder. The force of the bullet spun him around counter-clockwise, like a toy. He crashed against a stack of cement bags, splitting several sacks. He collapsed face down onto the desert sand. The shock of the wound almost drove Nolen unconscious, but his desire to live was stronger.

Fight to survive! his psyche commanded.

Fight to survive!

In an effort which seemed to last an eternity, Nolen struggled to his knees, using the steel rod for support. He leaned the right side of his body against the bags and gazed at his torn, bleeding shoulder.

Cement dust from the busted bags covered his chest and face, searing his eyes. Each time he sucked in jagged gasps of air, or licked his dry lips, the dust tasted bitter. The Texan's evil, smiling face leaned near, as a distant voice threatened, "See you in Hell, Asshole."

Behind Boodan, Maida shouted, "Get away from him! Leave him alone!" She had ignored Nolen's orders. After Luke had tumbled from the lumber stack, she had recovered his dropped rifle. At first, she had cautiously moved through the storage yard, attempting to find Nolen to give him the weapon. When she heard Boodan and Nolen arguing, she ran to the end of the alley, then stepped into the open. Now, off balance, she quickly raised the heavy rifle to her shoulder and fired. The kick from the large-bore weapon unexpectedly knocked her off her feet.

The ineffective shot diverted Boodan's attention to the new threat. "You little bitch!"

Fight to survive!

Nolen grabbed a handful of cement, and tossed it into Boodan's partially turned face. The caustic powder splashed over his enemy's one good eye, burning and blinding him. Tapping his remaining strength, Nolen struggled to his feet, stumbled backward several steps, and again fell to the ground. A hot nauseating pain gnawed at his chest, sapping his strength.

"Ahh!" the Cajun screamed and sightlessly fired his rifle three times at the scurrying woman. Infuriated, Boodan spun back around, to kill Nolen. His weapon exploded two times, sending bullets into the spot where Nolen had been leaning seconds earlier, blasting apart two additional cement bags.

To no avail, Boodan furiously rubbed his eye with his shirt sleeve. Viciously, he fired several times, again in Maida's direction, as she crashed out of the alley.

Boodan turned back toward Nolen. "Where the hell are you? I want to feel the blood drippin' out of your ugly body!"

Nolen held his breath, as he watched the blind menace turning and twitching in the moonlight, a mere eight feet away, stomping with his foot to locate Nolen. Quietly, carefully, Nolen picked up a stone. He tossed it across the alley, against a barrel.

Boodan whirled toward the noise, firing twice, blowing gaping holes in the metal container. He continued pulling the trigger until, at last, his rifle was empty.

"You missed," Nolen coughed.

Boodan threw the rifle down and pulled a silver bladed knife from his boot sheath. "I'll cut your heart into pieces," he yelled, as he charged forward, following the sound of Nolen's voice.

Nolen jerked up one end of the reinforcement rod, impaling his attacker through the chest. Boodan slid down the metal bar, slamming into the miner.

Nolen floated in and out of consciousness. Gradually, he became aware that he was being lifted into a helicopter. Those around him were shouting commands as they secured the gurney and bottle of plasma within the aircraft. The helicopter leaped into the air, then sped toward the Susanville hospital.

He struggled to focus his vision. Slowly, he recognized Maida and Father Gabe. "Maida, you're alive! I...I was afraid Boodan shot you. I was so angry." He struggled to grasp Maida's hand. "Pretty lady, I love you. I...I...want to marry you."

"This is not the time to propose," Maida said, stroking his face.

"I've made a terrible mistake. Get a lawyer. I must change my will...half my mine shares go to you...half to Hilda. I promised Digger I would take care of Hilda...I promised," Nolen babbled.

"Don't talk as if you're going to die!" Maida ordered.

"Maida, I thought it was finally over. That...I could be with you. But someone will always be after the gold."

"Easy, Nolen. You're safe now. You killed both attackers. Right after you stuck that evil, one-eyed man, I got Jake. He carried you back to the headquarters, where the town doctor immediately treated your wounds. Jake's here. He will ensure his men guard you during your recovery."

"It hurts, Maida. Please, don't leave me. Don't ever leave."

"I'll stay with you, Darling. Lie still. The medication the doctor gave you will take effect soon. The pain will subside."

Nolen kept repeating, "Fight to survive. Fight to survive."

Father Gabe looked at the doctor, who was taking Nolen's pulse. "Should I prepare him for his last rites?"

"No. He'll make it," the doctor answered. "I've handled hundreds of shooting victims, when I worked in emergency rooms in San Francisco."

Maida looked at the doctor. "What did you say?"

"Miss Collins, you have little need for concern. The bullet did not rupture a lung, nor sever an artery. It only broke his shoulder. He did lose considerable blood. But he received plasma within minutes of his injury, preventing him from going into shock. And the best thing going for him, is that he's a fighter. He'll stabilize soon. Don't worry. He won't die."

"Thank God!" Maida whispered, wiping tears from her cheeks. "Thank God."

The End

Express your opinion about the book to the author:

E-mail: ed@goldrush2000.net

Tell a friend!

Now that you enjoyed *Gold Rush 2000*,
please encourage a friend to buy a copy, or
purchase one for them as a gift.

Easy Order from our toll free number:
1-(888) 773-9769

Or from a book store by requesting the
book by title or ISBN.
ISBN: 0-9668447-3-4

Or send your request to:
California Coast Publishing
17595 Vierra Canyon Road, Suite 407
Salinas CA 93907